Millimeter-Wave Antennas for 5G

Millimeter-Wave Antennas for 5G

Guest Editor

Minmin Mao

MDPI

Basel • Beijing • Wuhan • Barcelona • Belgrade • Novi Sad • Cluj • Manchester

Guest Editor
Minmin Mao
College of Electronic
Information and Engineering
Hangzhou Dianzi University
Hangzhou
China

Editorial Office
MDPI AG
Grosspeteranlage 5
4052 Basel, Switzerland

This is a reprint of the Special Issue, published open access by the journal *Sensors* (ISSN 1424-8220), freely accessible at: https://www.mdpi.com/journal/sensors/special_issues/DZ8Q514XMO.

For citation purposes, cite each article independently as indicated on the article page online and as indicated below:

Lastname, A.A.; Lastname, B.B. Article Title. *Journal Name* **Year**, *Volume Number*, Page Range.

ISBN 978-3-7258-5417-2 (Hbk)
ISBN 978-3-7258-5418-9 (PDF)
https://doi.org/10.3390/books978-3-7258-5418-9

Contents

sensors MDPI

Article

Frequency-Selective Surface-Based MIMO Antenna Array for 5G Millimeter-Wave Applications

Iftikhar Ud Din [1], Mohammad Alibakhshikenari [2,*], Bal S. Virdee [3], Renu Karthick Rajaguru Jayanthi [3], Sadiq Ullah [1], Salahuddin Khan [4], Chan Hwang See [5], Lukasz Golunski [6,*] and Slawomir Koziel [6,7,*]

1 Telecommunication Engineering Department, University of Engineering and Technology, Mardan 23200, Pakistan; iftikharuddin114@gmail.com (I.U.D.); sadiqullah@uetmardan.edu.pk (S.U.)
2 Department of Signal Theory and Communications, Universidad Carlos III de Madrid, 28911 Leganés, Madrid, Spain
3 Center for Communications Technology, London Metropolitan University, London N7 8DB, UK; b.virdee@londonmet.ac.uk (B.S.V.); r.k.r.jayanthi@londonmet.ac.uk (R.K.R.J.)
4 College of Engineering, King Saud University, P.O. Box 800, Riyadh 11421, Saudi Arabia; drskhan@ksu.edu.sa
5 School of Computing, Engineering and the Built Environment, Edinburgh Napier University, 10 Colinton Rd., Edinburgh EH10 5DT, UK; c.see@napier.ac.uk
6 Faculty of Electronics, Telecommunications and Informatics, Gdansk University of Technology, 80-233 Gdansk, Poland
7 Engineering Optimization & Modeling Center, Reykjavik University, 101 Reykjavik, Iceland
* Correspondence: mohammad.alibakhshikenari@uc3m.es (M.A.); lukgolun@pg.edu.pl (L.G.); koziel@ru.is (S.K.)

Abstract: In this paper, a radiating element consisting of a modified circular patch is proposed for MIMO arrays for 5G millimeter-wave applications. The radiating elements in the proposed 2×2 MIMO antenna array are orthogonally configured relative to each other to mitigate mutual coupling that would otherwise degrade the performance of the MIMO system. The MIMO array was fabricated on Rogers RT/Duroid high-frequency substrate with a dielectric constant of 2.2, a thickness of 0.8 mm, and a loss tangent of 0.0009. The individual antenna in the array has a measured impedance bandwidth of 1.6 GHz from 27.25 to 28.85 GHz for $S_{11} \leq -10$ dB, and the MIMO array has a gain of 7.2 dBi at 28 GHz with inter radiator isolation greater than 26 dB. The gain of the MIMO array was increased by introducing frequency-selective surface (FSS) consisting of 7×7 array of unit cells comprising rectangular C-shaped resonators, with one embedded inside the other with a central crisscross slotted patch. With the FSS, the gain of the MIMO array increased to 8.6 dBi at 28 GHz. The radiation from the array is directional and perpendicular to the plain of the MIMO array. Owing to the low coupling between the radiating elements in the MIMO array, its Envelope Correlation Coefficient (ECC) is less than 0.002, and its diversity gain (DG) is better than 9.99 dB in the 5G operating band centered at 28 GHz between 26.5 GHz and 29.5 GHz.

Keywords: MIMO antenna; millimeter-wave (mm wave) region; wide bandwidth; mutual coupling reduction; frequency-selective surface; fifth generation (5G)

1. Introduction

The development of wireless communication networks has indeed been a significant modern uprising in the telecom industry. Wireless communication allows for the transmission of voice, data, and multimedia over the airwaves without the need for physical wired connections [1]. Between 2015 and 2020, global mobile data traffic saw a significant surge. According to various reports, the compound annual growth rate (CAGR) of mobile data traffic during this period was around 45%. This exponential increase was primarily driven by the growing number of smartphones, tablets, and other connected devices, as well as the rise of mobile applications and video streaming services. To meet the growing demand for wireless data, regulatory bodies and industry stakeholders continually work on allocating additional spectrum resources. Spectrum auctions and reallocations are conducted to ensure that a sufficient spectrum is available for mobile communication operators to deploy and expand their networks. To meet the data requirement challenges of wireless communications, 5G technology offers lower latency and improved spectral efficiency compared to previous generations. It employs advanced techniques such as massive multiple-input and multiple-output (MIMO), beamforming, and dynamic spectrum sharing to enhance capacity and improve spectral efficiency [2].

The 5G networks utilize both sub-6 GHz and millimeter-wave spectrum to provide enhanced performance and capabilities. Sub-6 GHz spectrum refers to the frequency bands below 6 GHz, including bands such as 600 MHz, 700 MHz, 2.5 GHz, 3.5 GHz, and 5 GHz [3,4]. The sub-6 GHz spectrum offers better coverage and penetration through buildings and obstacles compared to higher-frequency bands. It is well suited for providing wide-area coverage and supporting applications that require broader coverage, such as mobile broadband services. The millimeter-wave spectrum operates in higher-frequency bands, typically between 24 GHz and 100 GHz [5–9]. These frequencies offer significantly higher data transfer rates but have a limited range and are more susceptible to signal attenuation from obstacles like buildings and foliage. The millimeter spectrum is particularly well suited for providing high-capacity and low-latency connections in dense urban areas and specific hotspot locations. It enables the delivery of ultra-fast speeds and supports bandwidth-demanding applications such as virtual reality, augmented reality, and high-definition video streaming. However, millimeter-wave transmissions are affected by atmospheric attenuation, such as fog, rain, and snow. To overcome the challenges posed by atmospheric attenuation in millimeter-wave communications, high gain and high directive antennas are often used. These antennas focus the transmitted signal in a specific direction, increasing the signal strength and improving the overall link quality.

Multiple-input multiple-output (MIMO) antenna systems play a vital role in modern and upcoming mobile communication technologies. MIMO technology enables the use of multiple antennas at both the transmitter and receiver, working simultaneously to improve link reliability, enhance channel capacity, and achieve higher throughput in terms of gigabits per second (Gbps) [10–12]. The design of MIMO antenna arrays comes with several challenges that need to be addressed to ensure optimal performance and reliability. Some of the key challenges in designing MIMO antenna arrays include reduction in mutual coupling, the spatial arrangement of radiating elements, operational bandwidth, complexity and cost, physical size, and form factor. Mutual coupling occurs when the presence of one antenna affects the performance of other nearby antennas in the array. This coupling can lead to interference between antennas, reducing the overall system performance. Managing mutual coupling is crucial in MIMO antenna design to maintain antenna isolation and minimize the impact on signal quality. The spatial arrangement and placement of antenna elements in the array can significantly affect the performance of the MIMO system. Determining the optimal placement to achieve desired radiation patterns, coverage, and mutual coupling reduction is a complex task that requires careful analysis and optimization. MIMO antenna arrays need to operate across multiple frequency bands or wide bandwidths to support various communication standards and requirements. Ensuring consistent performance and impedance matching over a broad frequency range is a challenge in MIMO antenna design. As the number of antennas in the array increases, so does the complexity and cost of the MIMO system. Each antenna element requires its own RF chain, which adds complexity to the system design and increases hardware costs. Managing the complexity and cost while achieving the desired performance is a challenge in MIMO antenna array design. The physical size and form factor of MIMO antenna arrays are essential considerations, especially in mobile devices or small form factor applications. Balancing the desired antenna performance with space limitations can be a challenge, requiring innovative design techniques such as compact antenna structures and integration with other components.

The authors in [9] reported a four-element MIMO antenna array operating at 28 GHz with a gain of 5.42 dBi. The MIMO has an overall size 30 × 30 × 1.575 mm^3. In [13], the authors reported a 5G MIMO array that has a maximum gain of 7.1 dBi at 27 GHz and operates across 25.5–29.6 GHz. The MIMO antenna in [14] is shown to achieve a maximum gain of 8.3 dBi with a low isolation of 17 dB. Reported in [8] is a planar 28 GHz MIMO antenna inspired by a helix structure. The MIMO antenna has an impedance bandwidth of 3.89 GHz between 26.25 and 30.14 GHz, with an end-fire radiation pattern that has a maximum measured gain of 5.83 dBi. In [15], the authors reported a 28 GHz MIMO antenna for 5G applications where the radiating elements consists of conjoined triplet circular-shaped rings overlapped by conjoined two larger circular-shaped rings. The MIMOI array has a measured gain of 5.5 dBi. Reported in [16] is a 4 × 4 MIMO Dielectric Resonator Antenna (DRA) for 5G applications. DRA has an operating range from 26.71 GHz to 28.91 GHz with an isolation of 29 dB at 28 GHz, and a maximum gain of 7 dBi. The MIMO antenna reported in [17] has a Defected Ground Structure (DGS). The MIMO antenna is designed at 26.414 GHz and has a −6 dB bandwidth of 3.5346 GHz. The maximum gain of the MIMO antenna is limited to 6.22 dBi. Described in [18] is a two-element MIMO antenna for millimeter-wave applications at 28 GHz. The MIMO elements are arranged parallel to each other and placed between the elements is a metamaterial slab, which is shown to improve inter element isolation by 64 dB at 28 GHz. The maximum gain of the MIMO antenna is 8.75 dB at 28 GHz.

This paper describes the results of an investigation of improving the gain and inter radiation element isolation of a 2 × 2 MIMO antenna array for 5G millimeter-wave applications. To minimize the mutual coupling between the radiating elements in the MIMO array, the radiating elements are spatially arranged to be orthogonal with respect to each other. The proposed MIMO array has an impedance bandwidth of 1.6 GHz, a gain of 7.2 dBi, and an inter radiator isolation greater than 26 dB. It is shown here that the gain of the MIMO array can be enhanced by placing a frequency-selective surface (FSS) over the array. With the FSS, the gain achieved at 28 GHz is 8.6 dBi. The planar MIMO antenna array is compact and of a simple construction that facilitates easy integration in 5G wireless communication systems.

2. Antenna Design Procedure

The antenna configuration is based on a standard circular patch antenna that is edge-fed. The circular antenna is modified through four steps, depicted in Figure 1. In the first step, the circular patch antenna is designed. The initial radius (R) of the circular patch antenna in cm can be calculated using the following expression [19].

$$R = \frac{F}{\sqrt{1 + \frac{2h}{\pi \varepsilon_r F}\left(\ln\left(\frac{\pi F}{2h}\right) + 1.7726\right)}} \tag{1}$$

where $F = 8.791 \times 10^9 / f\sqrt{\varepsilon_r}$, h is the height of the substrate in cm, ε_r is the dielectric constant of the substrate, and f is the resonant frequency of the patch in Hz. The initial circular patch antenna was designed at 28 GHz based on Equation (1) using Roger's RT/Duroid 5880 substrate with a dielectric constant of 2.2, a thickness of 0.8 mm, and a loss tangent of 0.0009. The overall antenna size is 18 × 19 × 0.8 mm^3. Figure 2 shows the physical parameters defining the antenna. Figure 3 shows the simulated reflection coefficient response of the antenna corresponding to each step using CST Microwave Studio. The presence of a ground plane rectangular slot improves the impedance bandwidth of the patch antenna as it introduces additional resonant modes. Semicircular sections are cut out of the circular patch to center the resonant frequency at 28 GHz. The antenna resonates at the prescribed frequency of 28 GHz and has an impedance bandwidth of 1.6 GHz from 27.25 to 28.85 GHz for $S_{11} \le -10$ dB. The physical parameter dimensions of the antenna structure are listed in Table 1.

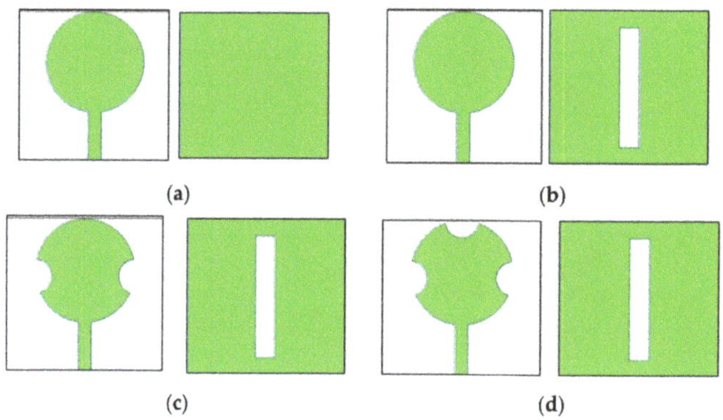

Figure 1. Evolution of the proposed radiating patch antenna. (**a**) Step #1. (**b**) Step #2. (**c**) Step #3. (**d**) Step #4.

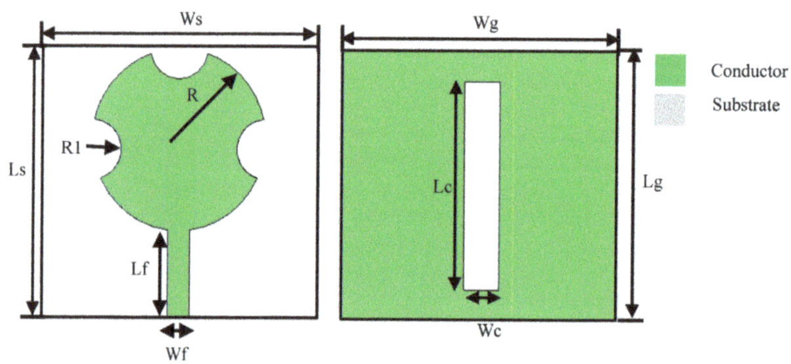

Figure 2. Front and back view of the proposed antenna.

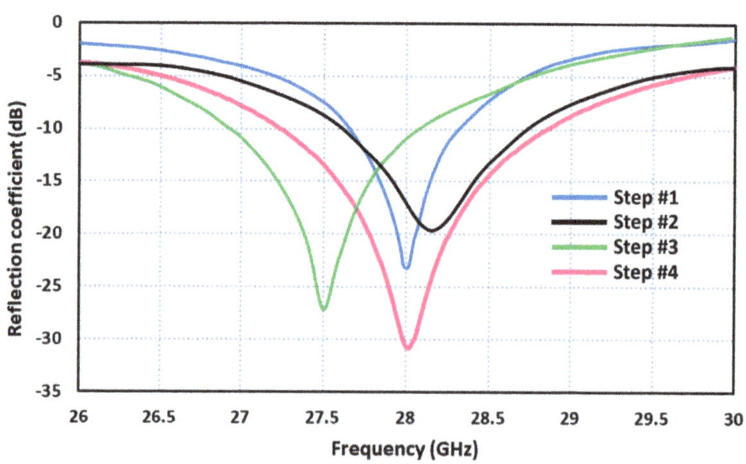

Figure 3. Reflection coefficient response of the modified circular patch antenna.

Table 1. Design parameters of modified circular patch antenna.

Symbol	Dimensions (mm)
L_s	18
L_g	18
L_f	6
W_f	1.5
R	6
R1	2
Lc	13.9
Wc	2.4
W_s	19
W_g	19

3. Effect of Ground Plane Rectangular Slot

The effect of the ground plane rectangular slot on the antenna's performance was analyzed in detail. Figure 4 shows how the dimensions of the slot affect the performance of the antenna. The parametric study shows that as the length of the slot is decreased, the resonance frequency of the antenna decreases, and the impedance bandwidth deteriorates significantly. A decrease in the slot width from 2.4 mm to 2 mm decreases the resonance frequency by 400 MHz with virtually no effect on the impedance bandwidth of the antenna. This property can be used to finetune the resonance frequency of the antenna. However, a further decrease in the width causes the impedance bandwidth to deteriorate significantly. The salient performance parameters of the antenna for various slot lengths and widths are listed in Table 2. The widest impedance bandwidth of 2 GHz is obtained for a slot length and width of 13.9 mm and 2.4 mm, respectively.

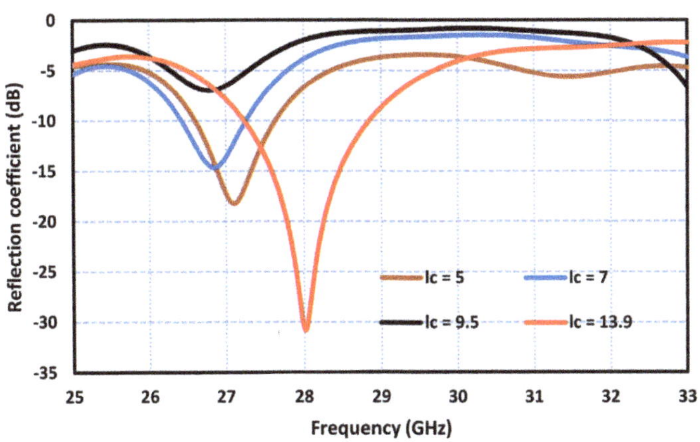

(a)

Figure 4. *Cont.*

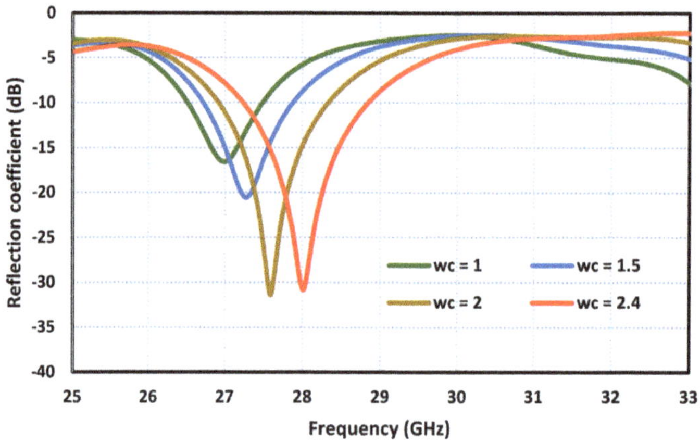

(**b**)

Figure 4. The effect on the reflection coefficient response of the antenna (**a**) by the slot length (Lc) and (**b**) by the slot width. Units of the length and width are in millimeters.

Table 2. Dimensions of ground plane rectangular slot used in parametric study.

Slot Length (Lc) (mm)	Slot Width (Wc) (mm)	Freq. Range (GHz)	Impedance BW (GHz)
5	1	26.6–27.6	1
7	1.5	26.4–27.2	0.8
9.5	2	33.2–34	0.8
13.9	2.4	27–29	2

4. MIMO Antenna Array

The modified circular patch antenna was used in the design of a 2 × 2 MIMO antenna array operating at 28 GHz. The individual antennas in the array were spatially arranged orthogonally with respect to each other, as shown in Figure 5 to improve the isolation between the individual antennas. This configuration also has the benefit of creating circular polarization. The MIMO antenna array was fabricated on Roger's RT/Duroid 5880 substrate with a dielectric constant of 2.2, a thickness of 0.8 mm, and a loss tangent of 0.0009. The overall dimension of the MIMO is 38 × 36 × 0.8 mm^3. The simulation results in Figure 6 show that at 28 GHz, the impedance matching at all four ports is better than −14 dB and the isolation between the radiating elements is better than 26 dB.

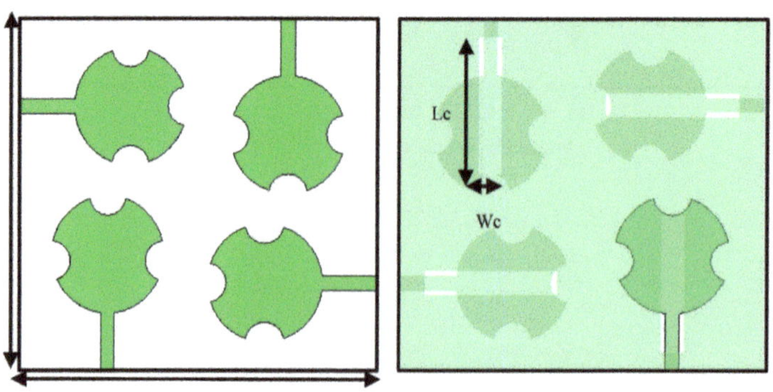

Figure 5. Front and back view of the proposed 2 × 2 MIMO antenna array.

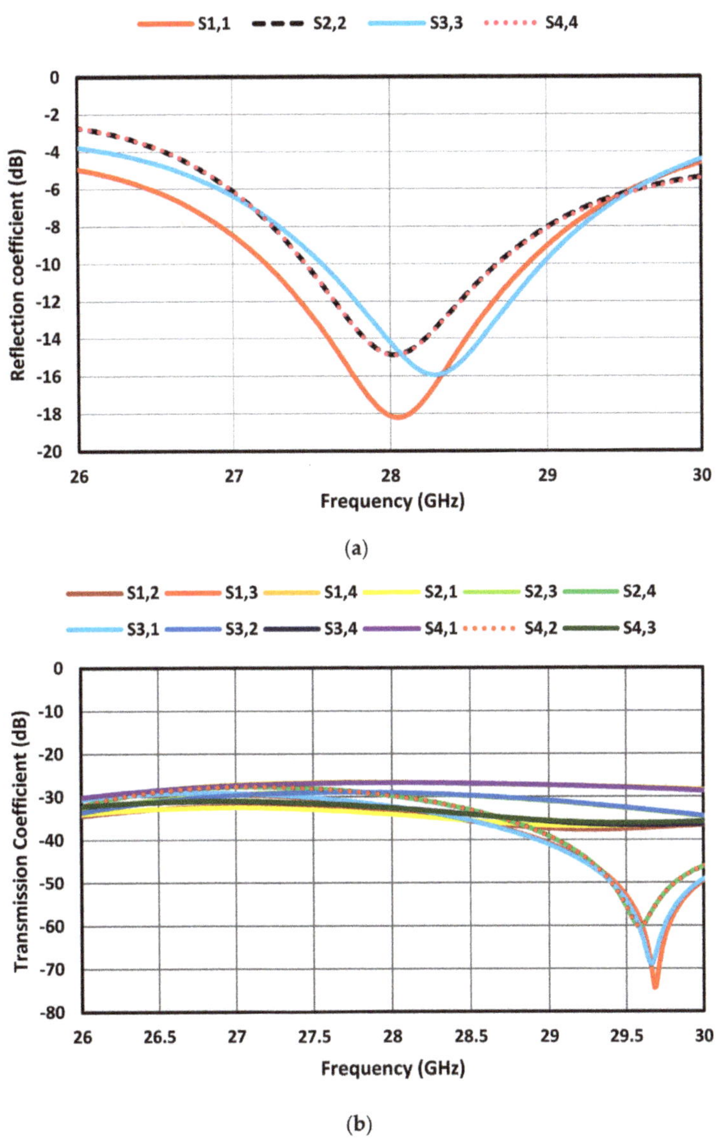

Figure 6. S-parameters of the proposed MIMO antenna array. (**a**) Reflection coefficient and (**b**) Transmission coefficient.

5. Frequency-Selective Surface Unit Cell

It is shown here that by locating the proposed frequency-selective surface (FSS), the proposed MIMO antenna array can reduce mutual coupling between the radiating elements, resulting in improved gain performance. This is because in an antenna array, mutual coupling occurs when the electromagnetic fields from one antenna interact with other antennas in the array. This interaction can lead to changes in the radiation pattern, impedance, and efficiency of the antennas, degrading the overall performance of the array. The FSS is used here as a bandpass structure, allowing signals to pass through the frequency range of interest and isolating the interaction of the antennas outside their frequency band. When electromagnetic (EM) waves incident on the FSS structure, they incite electric currents into the array elements. The level of coupling energy defines the magnitude of the produced currents. The generated currents also work as EM sources, and they create additional scattered fields. Incident EM fields combined with these scattered fields make up the resultant field in the surrounding of FSS. The operational theory of FSS-based structures has been explained by Munk in detail [20].

The steps taken to design the FSS unit cell are shown in Figure 7. The design starts with a C-shaped resonator. In the next step, a smaller C-shaped resonator is inserted inside the bigger C-shaped structure but facing the opposite direction. In the next step, a square patch is embedded inside the structure. In the final step, the central patch is divided with diagonal and vertical slots, as shown in Figure 7b. The dimensions of the FSS structure as defined in Figure 7b are listed in Table 3. Figure 7b also shows the boundary condition and port excitation used in CST Microwave Studio. The simulated reflection and transmission coefficient responses corresponding to each step are shown in Figure 7c. Step 4 of the FSS shows the impedance bandwidth for $S_{12} \leq -10$ dB is 2.3 GHz from 26.9 GHz to 29.2 GHz.

Table 3. Design parameters of the proposed FSS unit cell.

Parameters	Symbols	Value (mm)
Substrate length and width	Ls = Ws	3.7
Gap one	g1	0.24
Gap two	g2	0.25
Gap three	g3	0.27
Length of FSS	Lm	45
Width of FSS	Wm	45

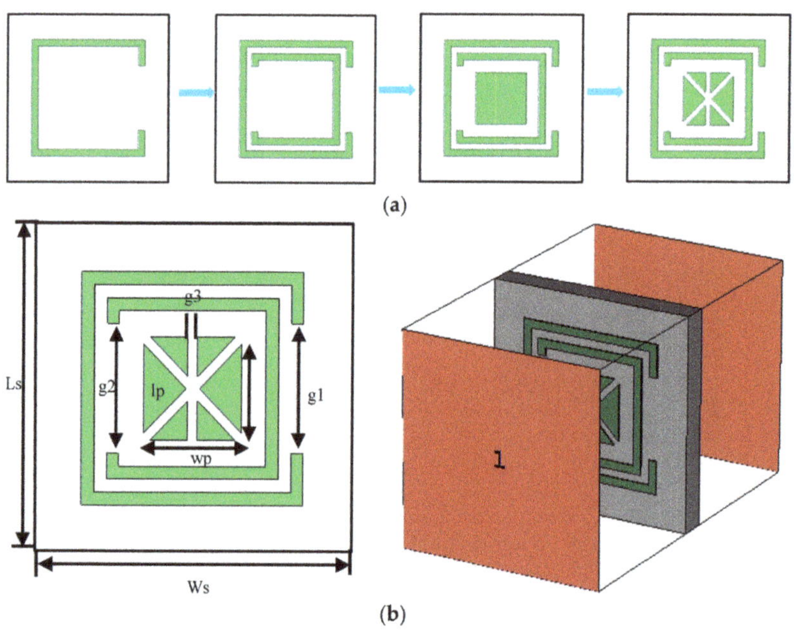

Figure 7. *Cont.*

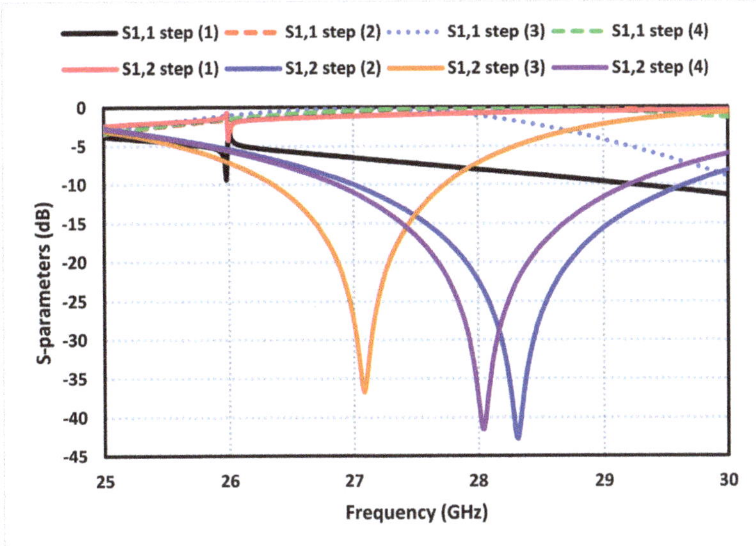

(c)

Figure 7. (**a**) Steps taken to create the proposed FSS unit cell. (**b**) Parameters defining the FSS unit cell and simulation excitation ports, and (**c**) S-parameter responses of the proposed FSS unit cell.

6. FSS-Loaded Antenna

The FSS, consisting of a 7 × 7 array of unit cell matrix, was used to enhance the performance of the antenna. This was achieved by locating the FSS structure over the MIMO antenna. The FSS structure allowed EM radiation to pass through that was within the operational frequency range of the MIMO antenna. The FSS structure was designed to have a bandpass between 27 GHz and 29 GHz. Figure 8 illustrates the mounting of the FSS with the proposed MIMO antenna. The gap between the FSS structure and the MIMO antenna was made to be an integer multiple of λg/2 for the reflected radiation to interfere constructively with the forward radiated waves. The FSS was fabricated on Roger's RT/Duroid 5880 substrate with a dielectric constant of 2.2, a thickness of 0.8 mm, and a loss tangent of 0.0009. The footprint of the FSS array is 45 × 45 mm^2.

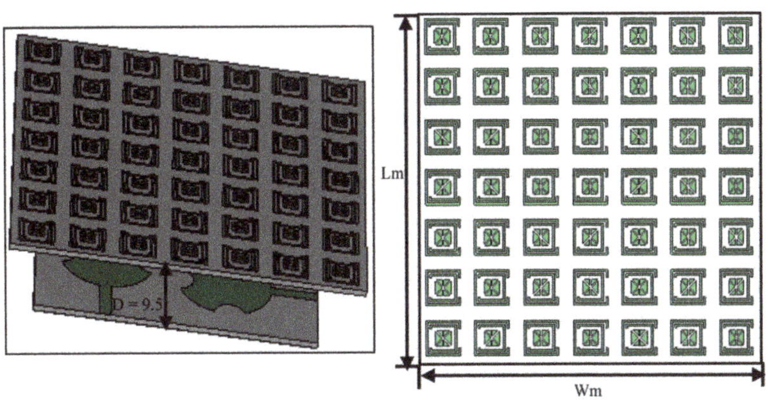

(a) (b)

Figure 8. (**a**) FSS array surface located under the MIMO antenna array, and (**b**) FSS reflector.

Figure 9 shows how the gain performance of the MIMO antenna was affected for various gaps between the FSS surface and the proposed MIMO antenna. At a gap of 7.2 mm corresponding to λg/2, the gain achieved at 28 GHz was 8.65 dBi; however, at a gap of 9.5 mm and 12 mm, the gain at 28 GHz was 8.1 dBi and 7.4 dBi, respectively. Therefore, a gap of 7.2 mm was used in the design of the MIMO antenna. The reflection coefficient at the four ports of the FSS-based MIMO antenna array is shown in Figure 10. It can be observed that the MIMO antenna array resonates at 28 GHz and its operational bandwidth for $S_{11} \leq -10$ dB is within the 5G mm wave spectrum between 26.5 GHz and 29.5 GHz.

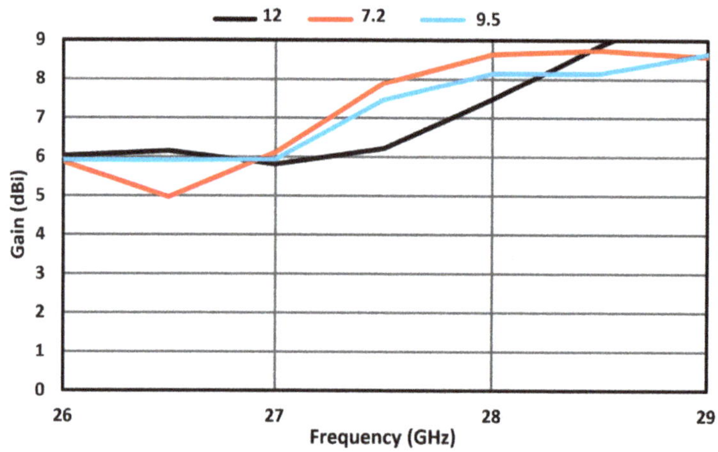

Figure 9. Gain of FSS-based MIMO antenna array at different gaps. Units are in millimeters.

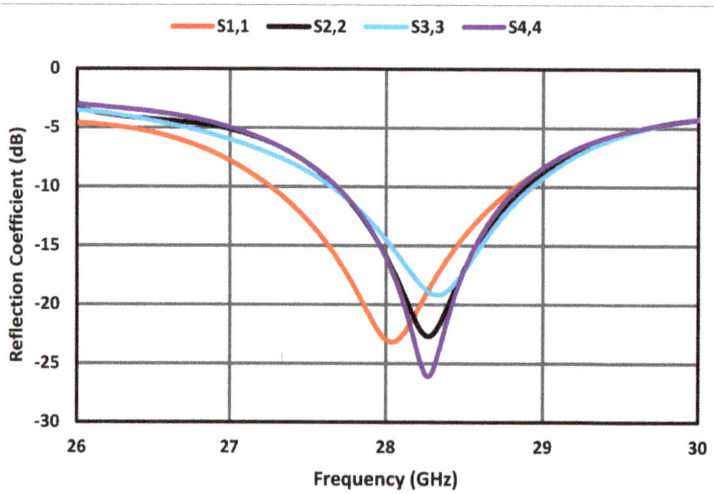

Figure 10. Reflection coefficient at the four ports of the FSS-based MIMO antenna array.

Figure 11 shows that the orthogonal arrangement of individual antennas in the proposed FSS-based MIMO array improves isolation between the radiating elements, which is better than 20 dB across a wide frequency range between 26 GHz and 30 GHz. This is because the correlation between the antennas is reduced. The correlation between antennas in an MIMO system is a measure of the similarity of their received signals. A high correlation between antennas can limit the system's ability to exploit the spatial diversity offered by MIMO, leading to a decrease in the achievable capacity gain. When the antennas are closely spaced or placed in a non-orthogonal configuration, they tend to experience higher mutual coupling and correlations. This is due to the unwanted electromagnetic interaction between adjacent antennas in the array. This coupling can result in signal interference, reduced antenna efficiency, and increased sensitivity to changes in the propagation environment.

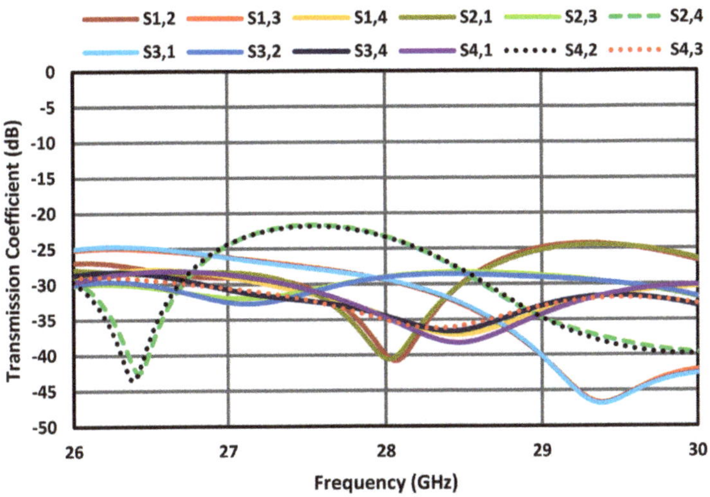

Figure 11. Transmission coefficients of the FSS-based MIMO antenna array.

7. Measured Results

The fabricated prototype of an FSS-based MIMO antenna array is shown in Figure 12. The proposed MIMO array was constructed on Roger's RT/Duroid 5880 substrate with a dielectric constant of 2.2, a thickness of 0.8 mm, and a loss tangent of 0.0009. The gap between the ground plane of the MIMO array and the FSS was 7.2 mm for the high gain performance at 28 GHz, as determined in Section 6. It was ensured that the power level and the phase of the excitation signals at the four ports were identical. The measured and simulated results are discussed in the subsequent sub-sections.

7.1. S-Parameters

The simulated and measured S-parameters of the FSS-based MIMO antenna are compared in Figure 13. The simulated impedance bandwidth, shown in Figure 13a, at port 1 is 1.65 GHz from 27.2 to 28.85 GHz, and at port 4 is 1.15 GHz from 27.7 GHz to 28.85 GHz. The measured impedance bandwidth at port 1 is 1.5 GHz from 27.1 to 28.6 GHz, and at port 4 is 0.95 GHz from 28 GHz to 28.95 GHz. The measured isolation between the four ports in Figure 13b is better than 20 dB. The discrepancy between the measured and simulated results of the impedance bandwidth and isolation between the radiating elements in the proposed MIMO antenna is attributed to several factors. These factors include manufacturing tolerances, unwanted antenna coupling, and inaccurate simulation models. Although an orthogonal arrangement is used to minimize mutual coupling, it is difficult to achieve complete isolation between radiating elements in practice. Some level of coupling might still exist, leading to deviations from the simulated results. Also, in simulations, certain assumptions and simplifications are made to facilitate computation. These assumptions might not fully capture all real-world complexities, leading to discrepancies between the simulated and measured results.

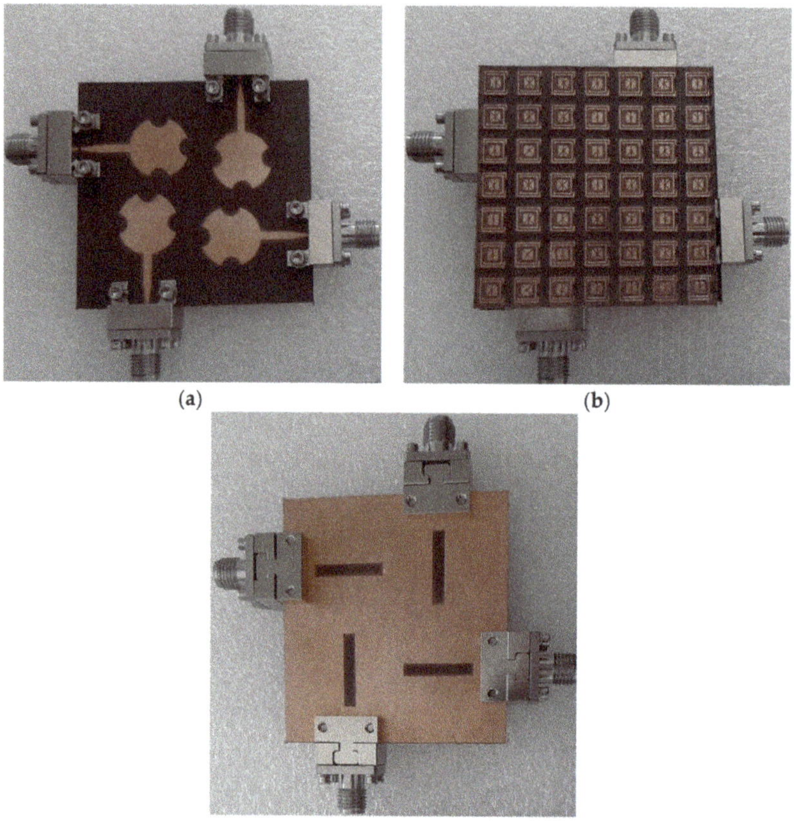

Figure 12. Fabricated prototype of the FSS-based MIMO antenna array, (**a**) front view, (**b**) FSS layer, and (**c**) back side view.

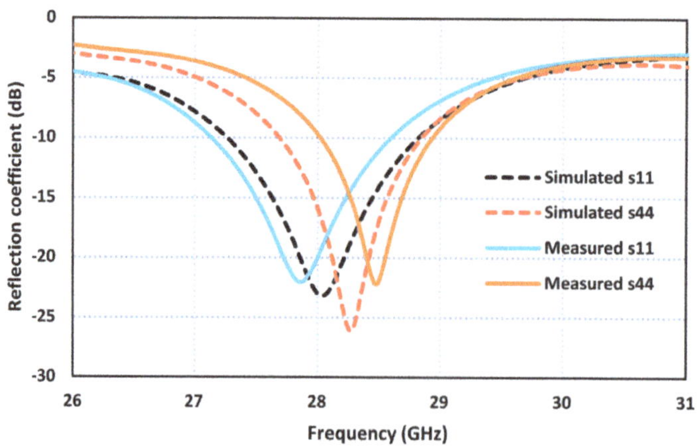

(**a**)

Figure 13. *Cont.*

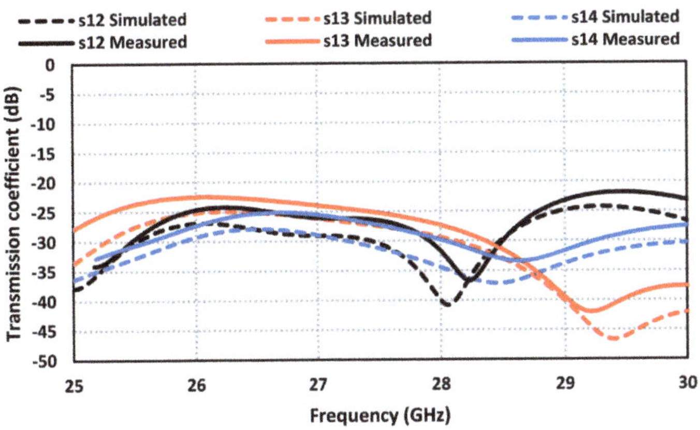

(b)

Figure 13. The measured and simulated results of the FSS-based MIMO antenna array. (**a**) Reflection coefficient response. (**b**) Transmission coefficient response.

7.2. 2D Radiation Patterns

The simulated and measured polar plots of FSS-based MIMO antenna at 28 GHz are shown in Figure 14a. The radiation patterns are shown for both E-plane and H-plane. The radiation from the MIMO antenna is directional and perpendicular to the plain of the antenna. The simulated gain at an angle of 0 degrees in the E- and H-planes is 8.4 dBi; however, the measured gain at this angle is 7.5 dBi. The measured radiation in the E- and H-planes is virtually identical over the 3-dB beamwidth of the MIMO antenna centered at 0 degrees. This confirms that circular polarization is maintained over the antenna's 3-dB beamwidth. Figure 14b shows how the gain of the array was affected by the inclusion of the FSS. Without FSS, the measured gain was 7.2 dBi, and with FSS, the gain increased to 8.6 dBi. The radiation efficiency of the MIMO antenna array at 28 GHz was measured to be 85%, and across its operational band, the efficiency was 80%. The discrepancy in the simulated and measured results as mentioned previously is attributed to manufacturing tolerances, unwanted antenna coupling, and inaccurate simulation models.

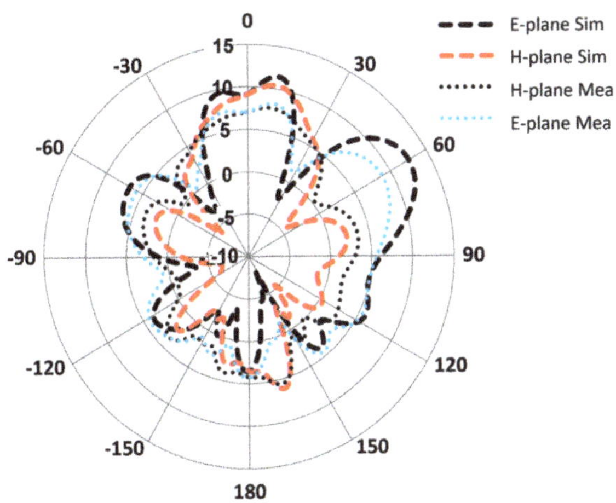

(a)

Figure 14. *Cont.*

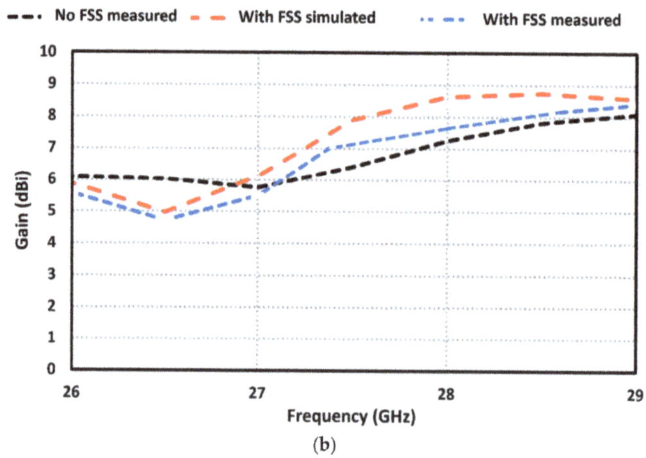

Figure 14. Measured and simulated plots of (**a**) radiation patterns and (**b**) gain of the FSS-based MIMO antenna array.

8. MIMO Performance Parameters

8.1. Envelope Correlation Coefficient

The Envelope Correlation Coefficient (ECC) is a metric used to quantify the correlation between two antennas in an MIMO system. A high correlation between antennas can limit the benefits of spatial diversity and reduce the potential capacity gain that MIMO offers. On the other hand, a low correlation is desirable because it allows the MIMO system to achieve higher data rates and improved link reliability. The value of ECC should be less than 0.5 to minimize coupling effects. The value of ECC can be estimated using the relation in [17].

$$ECC = \frac{|\iint 4\pi(M_i)(\theta,\phi)) \times (\, M_j(\theta,\phi))d\Omega|^2}{\iint 4\pi\left|(M_i(\theta,\phi))\right|^2 d\Omega \iint 4\pi\left|(M_j(\theta,\phi))\right|^2 d\Omega} \tag{2}$$

where $M_i\,(\theta,\phi)$ and $M_j\,(\theta,\phi)$ represent the radiation patterns when antennas i and j are excited, and the term Ω denotes the solid angle. The ECC of the proposed FSS-based MIMO antenna array as a function of frequency in Figure 15 shows that it is less than 0.002 in the 5G operating band at 28 GHz between 26.5 GHz and 29.5 GHz.

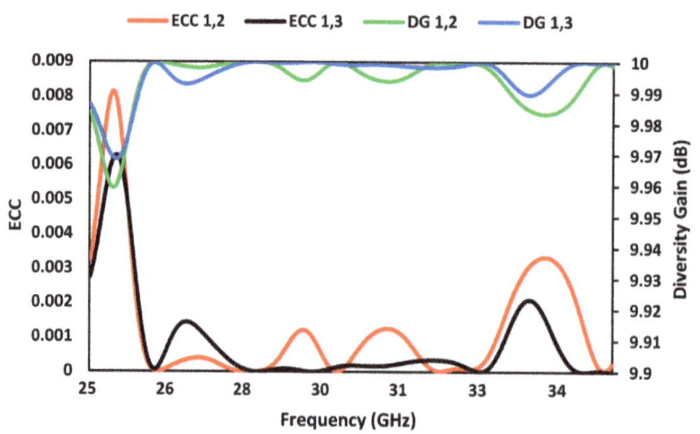

Figure 15. ECC and DG of the proposed FSS-based MIMO antenna array.

8.2. Diversity Gain

Diversity gain (DG) is the improvement in signal quality and system performance achieved by utilizing multiple antennas at both the transmitter and receiver. The diversity gain of MIMO antenna array can be calculated from the ECC using the following relation [17]:

$$DG = 10 \times \sqrt{(1 - ECC)} \quad (3)$$

The diversity gain of the proposed MIMO array in Figure 15 shows that it is better than 9.99 dB in the 5G operating band at 28 GHz between 26.5 GHz and 29.5 GHz.

The performance of the proposed MIMO antenna array is compared in Table 4 with other MIMO arrays reported in the literature. Compared to other four-port devices, the proposed MIMO array with FSS exhibits a higher gain and has the smallest ECC. The dimensions of the proposed MIMO array with FSS are comparable to other MIMO arrays cited in the table.

Table 4. Comparison with other reported MIMO antenna arrays.

Ref	f (GHz)	Ports	Dimension in (mm)	Min. Isolation (dB)	Gain (dBi)	ECC	Techniques
[9]	28	2	30 × 15 × 0.25	35.8	5.42	<0.005	DGS
[13]	27	4	30 × 30 × 1.575	30	7.1	<0.005	DGS
[14]	28	4	30 × 35 × 0.76	17	8.3	<0.010	DGS
[8]	28	2	15 × 25 × 0.203	30	5.8	<0.005	DGS
[15]	28	4	30 × 30 × 0.787	29	6.1	<0.160	DGS
[16]	28	4	20 × 40 × 1.6	29.34	7	<0.010	DRA
[17]	27	4	30 × 28 × 0.508	24	6.22	<0.050	DGS
[18]	28	2	18 × 36 × 0.8	64	8.75	<0.050	SRR/DGS
Proposed MIMO without FSS	28	4	38 × 36 × 0.8	26.31	7.2	<0.002	DGS
Proposed MIMO with FSS	28	4	45 × 45 × 0.8	23.31	8.6	<0.002	FSS/DGS

9. Conclusions

In this article, we have demonstrated that mutual coupling between neighboring antennas in an MIMO array can be reduced by spatially arranging the individual antennas orthogonally relative to each other. It is also shown that the impedance bandwidth of the antenna can be widened by defecting the ground plane with a rectangular slot. Moreover, it is shown that the gain of the MIMO array can be enhanced by locating a frequency-selective surface (FSS) over the MIMO array. The FSS unit cell was designed to allow the frequency band at 28 GHz. The gap between the MIMO array and FSS was adjusted to one-half guided wavelength to ensure constructive interference between the forward and FSS reflected electromagnetic waves. The frequency-selective surface consisted of a 7 × 7 matrix of FSS unit cells. The measured results show that the gain of the MIMO array is increased by including the FSS. The proposed FSS-based MIMO array has an impedance bandwidth between 27.2 GHz and 29 GHz with a total efficiency of 86% at 28 GHz. Moreover, its ECC is less than 0.002 and its diversity gain is better than 9.99 dB in the 5G operating band centered at 28 GHz between 26.5 GHz and 29.5 GHz.

Author Contributions: Conceptualization, I.U.D., M.A., B.S.V., R.K.R.J., S.U., S.K. (Salahuddin Khan), C.H.S., L.G. and S.K. (Slawomir Koziel); methodology, I.U.D., M.A., B.S.V. and R.K.R.J.; software, I.U.D., M.A., B.S.V. and R.K.R.J.; validation, I.U.D., M.A., B.S.V., R.K.R.J., S.U., S.K. (Salahuddin Khan), C.H.S., L.G. and S.K. (Slawomir Koziel); formal analysis, I.U.D., M.A., B.S.V., R.K.R.J., S.U., S.K. (Salahuddin Khan), C.H.S., L.G. and S.K. (Slawomir Koziel); investigation, I.U.D., M.A., B.S.V.,

R.K.R.J., S.U., S.K. (Salahuddin Khan), C.H.S., L.G. and S.K. (Slawomir Koziel); resources, I.U.D., M.A., B.S.V. and C.H.S.; data curation, I.U.D., M.A., B.S.V., R.K.R.J., C.H.S., L.G. and S.K. (Slawomir Koziel); writing—original draft preparation, I.U.D., M.A. and B.S.V.; writing—review and editing, I.U.D., M.A., B.S.V., R.K.R.J., S.U., S.K. (Salahuddin Khan), C.H.S., L.G. and S.K. (Slawomir Koziel); visualization, I.U.D., M.A., B.S.V., R.K.R.J., S.U., S.K. (Salahuddin Khan), C.H.S., L.G. and S.K. (Slawomir Koziel); supervision, M.A., B.S.V., C.H.S., L.G. and S.K. (Slawomir Koziel); project administration, M.A., B.S.V., C.H.S. and S.K. (Slawomir Koziel); funding acquisition, M.A., S.K. (Salahuddin Khan), L.G. and S.K. (Slawomir Koziel). All authors have read and agreed to the published version of the manuscript.

Funding: Mohammad Alibakhshikenari acknowledges support from the CONEX-Plus program funded by Universidad Carlos III de Madrid and the European Union's Horizon 2020 research and innovation program under the Marie Sklodowska-Curie grant agreement No. 801538. The authors also sincerely appreciate funding from Researchers Supporting Project number (RSP2023R58), King Saud University, Riyadh, Saudi Arabia. Besides above, the authors would like to thank Dassault Systemes, France, for making CST Microwave Studio available. Additionally, this work is partially supported by the Icelandic Centre for Research (RANNIS) Grant 239858 and by National Science Centre of Poland Grant 2020/37/B/ST7/01448.

Institutional Review Board Statement: Not applicable.

Informed Consent Statement: Not applicable.

Data Availability Statement: All data generated or analyzed during this study are included in this article. All of the figures, materials, and data within the manuscript are original and owned by authors.

Acknowledgments: Mohammad Alibakhshikenari acknowledges support from the CONEX-Plus program funded by Universidad Carlos III de Madrid and the European Union's Horizon 2020 research and innovation program under the Marie Sklodowska-Curie grant agreement No. 801538. The authors also sincerely appreciate funding from Researchers Supporting Project number (RSP2023R58), King Saud University, Riyadh, Saudi Arabia. Besides above, the authors would like to thank Dassault Systemes, France, for making CST Microwave Studio available. Additionally, this work is partially supported by the Icelandic Centre for Research (RANNIS) Grant 239858 and by National Science Centre of Poland Grant 2020/37/B/ST7/01448.

Conflicts of Interest: The authors declare no competing interest.

References

1. Abirami, M. A review of patch antenna design for 5G. In Proceedings of the 2017 IEEE International Conference on Electrical, Instrumentation and Communication Engineering (ICEICE), Karur, India, 27–28 April 2017; pp. 1–3.
2. Tariq, S.; Naqvi, S.I.; Hussain, N.; Amin, Y. A metasurface-based MIMO antenna for 5G millimeter-wave applications. *IEEE Access* **2021**, *9*, 51805–51817. [CrossRef]
3. Mohamed, H.A.; Edries, M.; Abdelghany, M.A.; Ibrahim, A.A. Millimeter-wave antenna with gain improvement utilizing reflection FSS for 5G networks. *IEEE Access* **2022**, *10*, 73601–73609. [CrossRef]
4. Jeong, M.J.; Hussain, N.; Park, J.W.; Park, S.G.; Rhee, S.Y.; Kim, N. Millimeter-wave microstrip patch antenna using vertically coupled split ring metaplate for gain enhancement. *Microw. Opt. Technol. Lett.* **2019**, *61*, 2360–2365. [CrossRef]
5. Przesmycki, R.; Bugaj, M.; Nowosielski, L. Broadband microstrip antenna for 5G wireless systems operating at 28 GHz. *Electronics* **2020**, *10*, 1. [CrossRef]
6. Kim, G.; Kim, S. Design and Analysis of dual polarized broadband microstrip patch antenna for 5G mmWave antenna module on FR4 substrate. *IEEE Access* **2021**, *9*, 64306–64316. [CrossRef]
7. Abdelaziz, A.; Mohamed, H.A.; Hamad, E.K.I. Applying characteristic mode analysis to systematically design of 5G logarithmic spiral MIMO patch antenna. *IEEE Access* **2021**, *9*, 156566–156580. [CrossRef]
8. Zahra, H.; Awan, W.A.; Ali, W.A.E.; Hussain, N.; Abbas, S.M.; Mukhopadhyay, S. A 28 GHz broadband helical inspired end-fire antenna and its MIMO configuration for 5G pattern diversity applications. *Electronics* **2021**, *10*, 405. [CrossRef]
9. Hussain, N.; Awan, W.A.; Ali, W.; Naqvi, S.I.; Zaidi, A.; Le, T.T. Compact wideband patch antenna and its MIMO configuration for 28 GHz applications. *AEU Int. J. Electron. Commun.* **2021**, *132*, 153612. [CrossRef]
10. Rahman, M.; NagshvarianJahromi, M.; Mirjavadi, S.S.; Hamouda, A.M. Compact UWB band-notched antenna with integrated Bluetooth for personal wireless communication and UWB applications. *Electronics* **2019**, *8*, 158. [CrossRef]
11. Sharawi, M.S. Printed multi-band MIMO antenna systems and their performance metrics. *IEEE Antennas Propag. Mag.* **2013**, *55*, 218–232. [CrossRef]

12. Ojaroudi Parchin, N.; Basherlou, H.J.; Alibakhshikenari, M.; Parchin, Y.O.; Al-Yasir, Y.I.A.; Abd-Alhameed, R.A.; Limiti, E. Mobile-phone antenna array with diamond-ring slot elements for 5G massive MIMO systems. *Electronics* **2019**, *8*, 521. [CrossRef]
13. Hussain, M.; Ali, E.M.; Jarchavi, S.M.R.; Zaidi, A.; Najam, A.I.; Alotaibi, A.A.; Althobaiti, A.; Ghoneim, S.S.M. Design and characterization of compact broadband antenna and its MIMO configuration for 28 GHz 5G applications. *Electronics* **2022**, *11*, 523. [CrossRef]
14. Khalid, M.; Naqvi, S.I.; Hussain, N.; Rahman, M.; Fawad; Mirjavadi, S.S.; Khan, M.J.; Amin, Y. 4-Port MIMO antenna with defected ground structure for 5G millimeter wave applications. *Electronics* **2020**, *9*, 71. [CrossRef]
15. Kamal, M.M.; Yang, S.; Ren, X.-C.; Altaf, A.; Kiani, S.H.; Anjum, M.R.; Iqbal, A.; Asif, M.; Saeed, S.I. Infinity shell shaped MIMO antenna array for mm-wave 5G applications. *Electronics* **2021**, *10*, 165. [CrossRef]
16. Murthy, N. Improved isolation metamaterial inspired mm-Wave MIMO dielectric resonator antenna for 5G application. *Prog. Electromagn. Res. C* **2020**, *100*, 247–261. [CrossRef]
17. Khalid, H.; Khalid, M.; Fatima, A.; Khalid, N. 2 × 2 MIMO antenna with defected ground structure for mm-wave 5G applications. In Proceedings of the 13th International Conference on Mathematics, Actuarial Science, Computer Science and Statistics (MACS), Karachi, Pakistan, 14–15 December 2019; pp. 1–6.
18. Din, I.U.; Ullah, S.; Mufti, N.; Ullah, R.; Kamal, B.; Ullah, R. Metamaterial-based highly isolated MIMO antenna system for 5G smartphone application. *Int. J. Commun. Syst.* **2023**, *36*, e5392. [CrossRef]
19. Balanis, C.A. *Handbook of Microstrip Antennas*; John Wiley and Sons: New York, NY, USA, 1982.
20. Munk, B.A. *Frequency Selective Surfaces: Theory and Design*; Wiley Online Library: Hoboken, NJ, USA, 2000; Volume 29.

sensors

MDPI

Article

A Compact Dual-Band Millimeter Wave Antenna for Smartwatch and IoT Applications with Link Budget Estimation

Parveez Shariff Bhadrvathi Ghouse [1], Pallavi R. Mane [1], Sangeetha Thankappan Sumangala [2], Vasanth Kumar Puttur [1], Sameena Pathan [2], Vikash Kumar Jhunjhunwala [3] and Tanweer Ali [1,*]

[1] Department of Electronics and Communication Engineering, Manipal Institute of Technology, Manipal Academy of Higher Education, Manipal 576104, India; parveez.bg@learner.manipal.edu (P.S.B.G.); palvi.mane@manipal.edu (P.R.M.); vasanth.puttur@manipal.edu (V.K.P.)

[2] Department of Information and Communication Technology, Manipal Institute of Technology, Manipal Academy of Higher Education, Manipal 576104, India; sangeetha.ts@manipal.edu (S.T.S.); sameena.bp@manipal.edu (S.P.)

[3] Department of Electrical and Electronics Engineering, Manipal Institute of Technology, Manipal Academy of Higher Education, Manipal 576104, India; vikas.kumar@manipal.edu

* Correspondence: tanweer.ali@manipal.edu

Abstract: Advancement in smartwatch sensors and connectivity features demands low latency communication with a wide bandwidth. ISM bands below 6 GHz are reaching a threshold. The millimeter-wave (mmWave) spectrum is the solution for future smartwatch applications. Therefore, a compact dual-band antenna operating at 25.5 and 38 GHz is presented here. The characteristics mode theory (CMT) aids the antenna design process by exciting Mode 1 and 2 as well as Mode 1–3 at their respective bands. In addition, the antenna structure generates two traverse modes, TM_{10} and TM_{02}, at the lower and higher frequency bands. The antenna measured a bandwidth (BW) of 1.5 (25–26.5 GHz) and 2.5 GHz (37–39.5 GHz) with a maximum gain of 7.4 and 7.3 dBi, respectively. The antenna performance within the watch case (stainless steel) showed a stable $|S_{11}|$ with a gain improvement of 9.9 and 10.9 dBi and a specific absorption rate (SAR) of 0.063 and 0.0206 W/kg, respectively, at the lower and higher bands. The link budget analysis for various rotation angles of the watch indicated that, for a link margin of 20 dB, the antenna can transmit/receive 1 Gbps of data. However, significant fading was noticed at certain angles due to the shadowing effect caused by the watch case itself. Nonetheless, the antenna has a workable bandwidth, a high gain, and a low SAR, making it suitable for smartwatch and IoT applications.

Keywords: dual-band; IoT; link margin; millimeter wave antenna; SAR; smartwatch

1. Introduction

Smartwatches (SWs) have become an integral part of our lives, as they are used for health and fitness parameter tracking, entertainment, phone calls, controlling other Internet of Things (IoT) devices, Global Positioning System (GPS) navigation, and so on [1]. In 2023 alone, it is estimated that there are 210.8 million SW users globally, which is expected to grow to 229.51 million users by 2027 [2]. With the rapid development in technology, SWs are loaded with sophisticated sensors to record physiological data such as respiratory [3], cardiac activity [4,5], and blood pressure [6]. The telemetry of these data from SW to the remote diagnostic center via the internet requires seamless connectivity. A precisely tailored antenna with a compact design, a low specific absorption rate (SAR), a good radiation pattern, a high gain, and an ease of embedding in SWs can bridge this gap.

The $0.472\lambda_0 \times 0.472\lambda_0$ (with λ_0 at 2.44 GHz) antenna can house a ground plane with four vertical metal rims [7]. This results in a low SAR of 1.24 and 1.44 W/kg for SW applications at 2.44 and 3.5 GHz, respectively. In [7], a metal ring-shaped antenna of 23 mm for bezel SW is proposed. The antenna efficiency and low electric-field distribution on the wrist were improved by placing a metal plate at the bottom. However, the actual bandwidth (BW) was 66 and 101 MHz at 1.55 and 2.4 GHz, respectively. As described in [8], multiple inverted L-shaped structures were etched on a $0.3\lambda_0 \times 0.35\lambda_0$ substrate operating at 0.88–0.96, 2.4–2.48, 3.3–50, and 5.1–5.8 GHz, achieving SAR in the range of 0.84–1.21 W/kg. In the later designs, the SW screen effect was not considered. Generally, the SW screen has a printed circuit board (PCB) with perfect electric conductor (PEC) cladding that supports the SW screen. Thus, embedding the antenna below the screen affects the radiation pattern. In [9], these effects were studied by considering flexible printed circuit (FPC) communication strips. The FPC strips carried signals to/from the PCB board to the SW screen. Two conductive tape-based structures were carved on a flexible printed circuit (FPC) to provide shielding. The length of these tapes was tuned to achieve resonances at 1.6, 2.4, 3.5, and 4.8 GHz with a structure dimension of $0.23\lambda_0 \times 0.28\lambda_0$.

The above-discussed structures have a large antenna profile with narrow bandwidths. The forthcoming SW is expected to support video calling and video streaming. Therefore, the millimeter wave spectrum that offers a wide bandwidth with low latency is the solution for personal area IoT applications [10]. In [11], a triple-band CPW monopole antenna was proposed for IoT applications with bandwidths of 22–30, 37–39, and 56.5–61 GHz. They achieved a SAR in the 0.9–0.62 W/kg range and a gain of 5.29, 7.49, and 9 dBi. Another design in [12] proposed a triple-band antenna for IoT applications. The structure with a dimension of $2.8\lambda_0 \times 2.35\lambda_0$ (with λ_0 at 28 GHz) combined two varied-length monopoles with a conducting stub at the bottom. The achieved bandwidths were 445 MHz, 657 MHz, and 5.14 GHz at 3.5, 5.8, and 28 GHz, and an antenna gain of 1.86, 2.55, and 4.41 dBi.

Very few designs exist that are suitable for SW IoT applications. Therefore, we designed a compact dual-band $0.51\lambda_0 \times 0.51\lambda_0$ (with λ_0 at 25.5 GHz) antenna with measured antenna bandwidths of 1.5 (25–26.5 GHz) and 2.5 GHz (37–39.5 GHz) and gains of 7.4 and 7.9 dBi, respectively. The antenna evolved through characteristic mode analysis (CMA), which was first proposed by Garbacz [13,14], providing a physical insight into an arbitrarily shaped antenna. Recently, CMA has been used to aid the design and configuration of MIMO, circular polarization, dual-polarization [15], and reconfigurable antennas [16,17]. The electromagnetic behavior of the antenna is mathematically modeled using weighted eigenvalue (λ_n), which depicts the number of significant modes that contribute to the resonance and provides information on the orthogonal electric field and surface current distribution of these modes [18]. It also provides information about the modes impedance characteristics, be they purely resistive, inductive, or capacitive. This information aids in tuning the antenna close to the ideal radiating conditions. The solution to λ_n conceptualizes the radiator's resonance, inductive, or capacitive mode [19], which can be altered by modifying the antenna structure.

The antenna performance was studied for SW applications by placing it inside a watch case with the material properties of stainless steel. The top of the watch case was covered with a reinforced glass of PTFE material. The proposed antenna with stainless steel at the bottom resulted in the lowest SAR of 0.063 and 0.0206 W/kg at respective bands. The SW and mobile phone link budget analysis was performed at 0.5 m for different watch angles. This showed satisfactory results with data supports of up to 1 Gbps. Therefore, the major contributions of this study are as follows:

1. the design and development of a compact dual-band antenna with the aid of characteristic mode theory to comprehend the resonant modes and their resultant radiation pattern;
2. an analysis of the performance of the proposed antenna for smartwatch applications by embedding it inside the watch case;
3. a study of the RF energy exposure to the human body caused by the proposed antenna with a watch case;
4. an estimation of the data handling capacity of the proposed antenna for a smartwatch at various angles in three-dimensional space.

2. Characteristic Mode-Based Antenna Methodology

The characteristic mode analysis (CMA) is a rapid development tool applied to the antenna development stages to comprehend its performance and radiation pattern. The structure was modified to achieve the desired results. CMA estimates the effective modes of an arbitrarily shaped antenna structure, where the first five modes are mostly found to be significant. However, modifying the structure and changing the feed position can generate undesired modes. In Stage 1 of the antenna design, an initial rectangular patch antenna was designed based on the design equations in [20] for a resonating frequency of 25.5 GHz.

From the above calculations, the approximate patch width and length were 4.65 and 3.718 mm, respectively. However, to retain compactness and to study the modified structure behavior, the patch width and length were chosen to be 5 and 3 mm in the first stage, as shown in Figure 1a. For CMA, a $0.51\lambda_0 \times 0.51\lambda_0$ lossless Roger 5880 substrate with a thickness of 0.25 mm and relative permittivity (ε_r) of 2.2 was selected. The electromagnetic simulation software CST v21 was used to analyze antenna development stages without port excitation. The antenna was designed to generate dual-band resonances at 25.5 and 38 GHz. Therefore, the CMA was performed separately for each of these frequencies. During CMA, the total surface current on the antenna structure of the patch and ground plane was estimated by considering the sum of the eigenvector (J_n) using Equation (1) [15,18].

$$J = \sum\nolimits_n \frac{V_n J_n}{1 + j\lambda_n} \tag{1}$$

where V_n is the model-excitation coefficient. The V_n along with model significance (MS) defines the significant modes at desired frequencies, where $MS = \left|\frac{1}{1+j\lambda_n}\right|$. In Figure 1b, the modal significance graph indicates that Mode 1 and 2 are significant at 25.5 GHz, Mode 1–3 are significant at 28 GHz with a wide bandwidth, and Mode 1–5 are significant at 38.5 GHz with a narrow bandwidth. The convergence of these modes eigenvalue (λ_n) to zero indicates that the modes are resonant. However, the modes with an eigenvalue $(\lambda_n) >$ or <0 are non-resonant inductive and capacitive modes. These modes store the energy in the form of electric and magnetic fields in the near-field region. The eigenvalue (λ_n) indicates that Mode 1–3 and 1–4 at 28 and 38.5 GHz are resonances, as shown in Figure 1c.

The characteristics angle (CA) is another feature of CMA, which represents the resonant and non-resonant modes by analyzing the phase lag between the electric field and surface current on the antenna structure. It is represented in terms of angle (α_n). The characteristics angle in Figure 1d depicts that Mode 1–3 from 27–28.5 GHz have zero eigenvalues, resulting in $\alpha_n = 180^\circ$, which means these modes are resonant modes with good impedance matching. Beyond this frequency, the Mode 1 and 2 surface current's phase lag from the electric field led to an inductive (non-resonant mode) mode. At 38.5 GHz, a poor impedance matching with a narrow bandwidth can be seen due to the summation of phase-lead and phase-lag of Mode 3 and 5 and Mode 2, respectively. The port excitation corroborates in the same manner, which indicates that, at 30 GHz, a good impedance matching |S11| of 37 dB with a bandwidth of 29.5–30.5 GHz is achieved, as shown in Figure 1e. However, at 38 GHz, due to inductive and capacitive Mode 2, 3, and 5, a poor impedance matching of 12 dB can be seen. Therefore, the CMA estimation for the first stage closely agrees with the port excitation results.

The sum of all modes of eigenvector currents (J_n) represents the total current on the antenna structure. In Figure 2, the eigenvector current (J_n) is concentrated over the antenna surface edges. Based on the J_n movement and direction, the respective modes characterize the radiation behavior. The eigenvector current and radiation pattern of the respective modes at 25.50 and 38 GHz are shown in Figure 2a,b, respectively. In Figure 2a, Mode 1 and 2 have in-phase vector currents (currents on the parallel edges are in the same direction) along the vertical/horizontal edges of the ground (black arrow) and radiator (red arrow). Mode 1 has an in-phase current along the vertical edges, causing a bidirectional radiation pattern with a null in the azimuthal plane at ϕ = 0°.

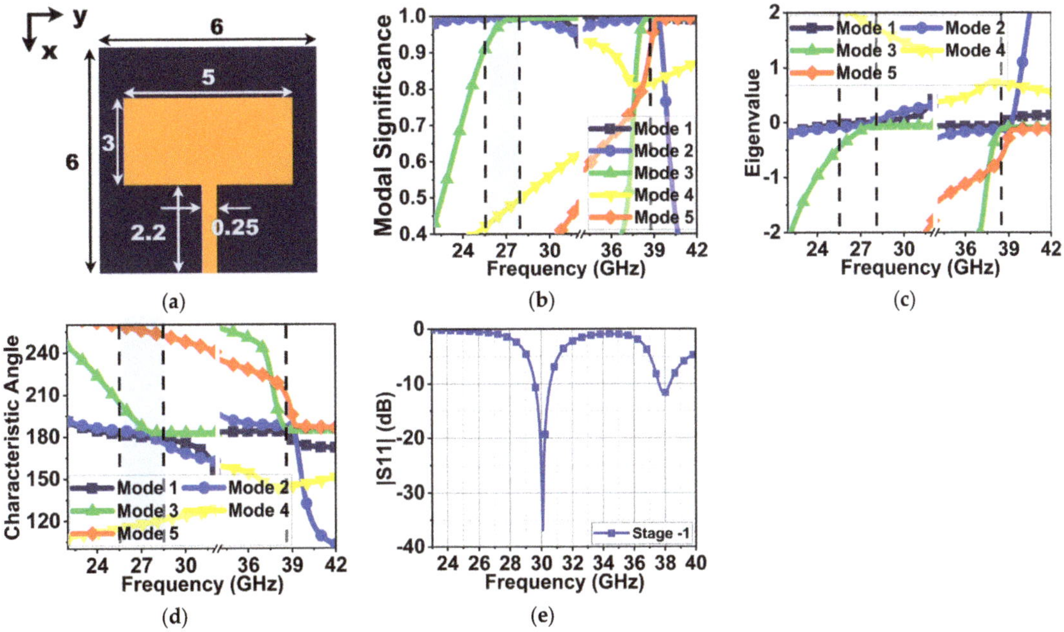

Figure 1. CMA of Mode 1–5 and reflection coefficients of the Stage 1 MS antenna. (**a**) Stage 1 antenna (dimensions in mm), (**b**) modal significance, (**c**) eigenvalues, (**d**) characteristic angle, and (**e**) reflection coefficient with port excitation.

Similarly, Mode 2 has an in-phase current at horizontal edges, providing a bidirectional radiation pattern with nulls in the azimuthal plane at ϕ = 90°. Mode 3–5 have an anti-phase eigenvector current (current in the opposite direction of the parallel edges) along the antenna edges. Thus, a quad/end-fire radiation pattern is formed, as seen in Figure 2a.

Similar effects can be observed in Figure 2b at 38 GHz but with the most current in the radiator. Mode 1 and 2 have an in-phase current on one set of parallel edges and an anti-phase current on the other, resulting in bidirectional quad radiation. However, Mode 3–5 have an anti-phase current, resulting in end-fire radiation.

As discussed above, the first resonance occurred at 30 GHz with good impedance matching and the second at 38 GHz with poor impedance matching. Further, in Stage 2, to decrease the first resonance to the expected frequency of 25.5 GHz and improve the impedance matching at 38 GHz, a semi-elliptical stub was added at the bottom of the radiator, as displayed in Figure 3a. From the CMA of Stage 2, the results of the MS in Figure 3b indicate that Mode 1–3 drifted from 30 to 28 GHz due to structure modification. However, at 25.5 GHz, only Mode 1 and 2 were found to be significant. On the other hand, Mode 1, 2, 3, and 5 drifted slightly higher, from 38.5 to 39 GHz. The eigenvalue of these modes converged to zero, as shown in Figure 3c, at respective frequencies, resulting in resonant modes with bandwidths of 24.1–26, 27–28, and 37.5–39 GHz. The characteristic angle shows a 180° phase lag between the surface current and the electric field at 25.5, 28, and 39 GHz of the resonant modes. However, the Mode 5 surface current led the electric field by 210° until 39 GHz, abruptly changing to a phase lag of 120° after 39 GHz. The Mode 4 surface current had a phase lag of 150° from 36 to 40 GHz. Due to these phase differences in Mode 4 and 5, the structure could not radiate complete energy at 39 GHz. As a result, an impedance mismatch occurred. The port excitation substantiated the CMA. The results indicate the occurrence of resonances at 28 and 38 GHz with impedance matching of 25 and 10.5 dB, as shown in Figure 3e.

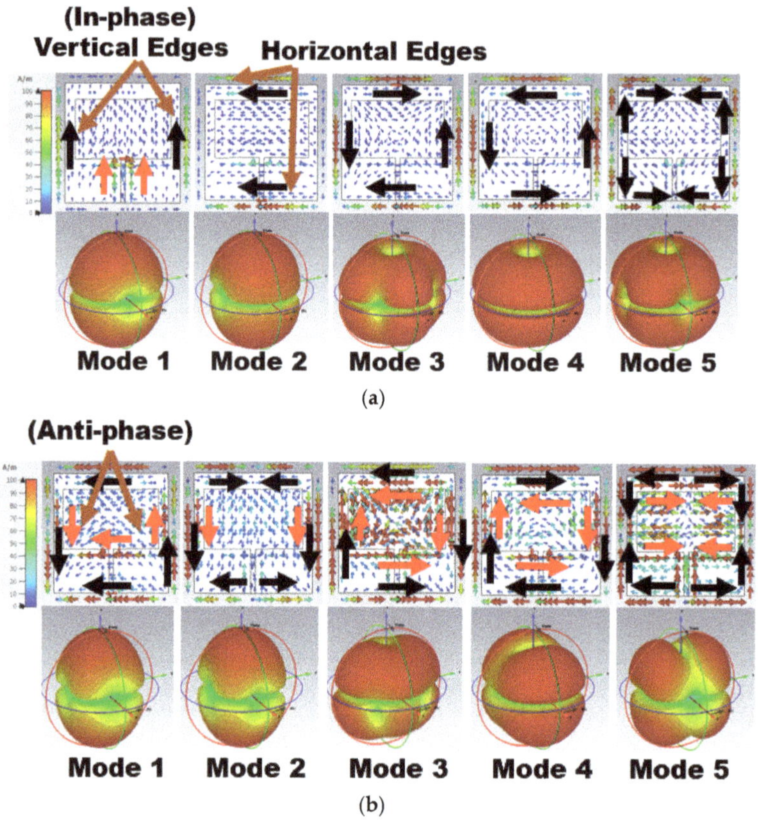

Figure 2. Eigenvector current J_n on the surface of Stage 1 and its respective radiation pattern at (**a**) 25.50 GHz and (**b**) 38 GHz.

Figure 4 represents the eigenvector current at 25.5 and 38 GHz. From the above eigenvalue and characteristic angle, only Mode 1 and 2 are resonant, but these are insufficient to generate the resonance at 25.5 GHz. This is evident from the eigenvector current in Figure 4a, as most of the current is in the ground plane rather than the radiator. At 38 GHz, Mode 1–3 have an in-phase current generating resonance, as shown in Figure 4b. However, an anti-phase current arose due to the non-resonant Mode 4 and 5, causing impedance matching.

Stage 3 with a semi-elliptical stub at the top of the radiator was introduced, as displayed in Figure 5a, to lower the resonance from 28 to 25.5 GHz and to improve the impedance matching at 38 GHz. This improved the significance of Mode 1–3 at 25.5 and 38 GHz, as shown in Figure 5b. The convergence of these modes to zero in the eigenvalue, as shown in Figure 5c, represents resonant modes (Mode 3 at 25.5 GHz is partially resonant) with bandwidths of 25–27 and 37–39 GHz. Mode 1–3 have an approximately 180° phase lag between the surface current and the electric field, as represented by the characteristics angle in Figure 5d. The modification to the structure significantly suppressed Mode 4 and 5 at 25.5 and 38 GHz. With the port excitation, dual-resonance can be seen in Figure 5e at 25.5 and 38 GHz with good impedance matching of 17.5 dB and 22.5 dB. It achieved a 25–25.9 GHz and 37.9–38.9 GHz bandwidth.

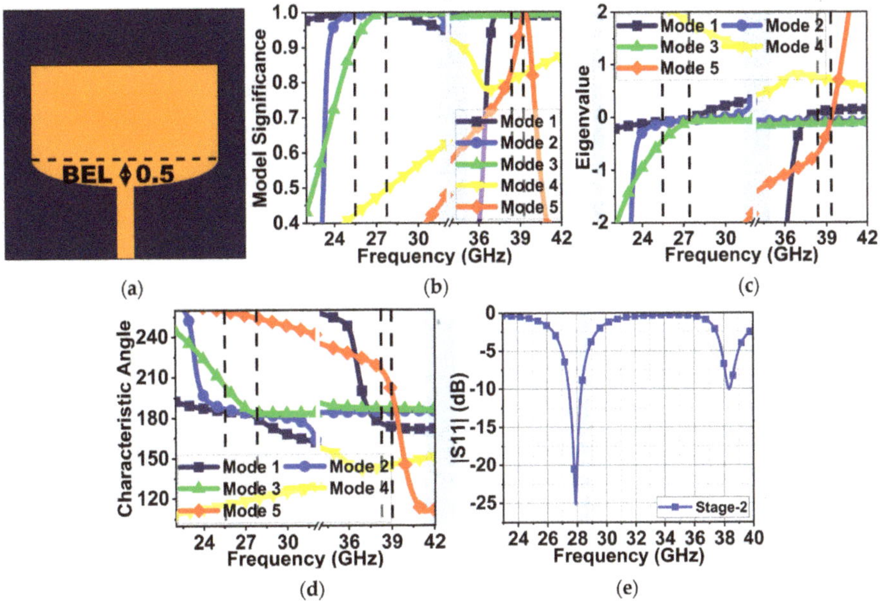

Figure 3. CMA of Mode 2–5 and reflection coefficients of the Stage-1 MS antenna. (**a**) Stage-1 antenna (dimensions in mm), (**b**) modal significance, (**c**) eigenvalues, (**d**) characteristic angle, and (**e**) reflection coefficient with port excitation.

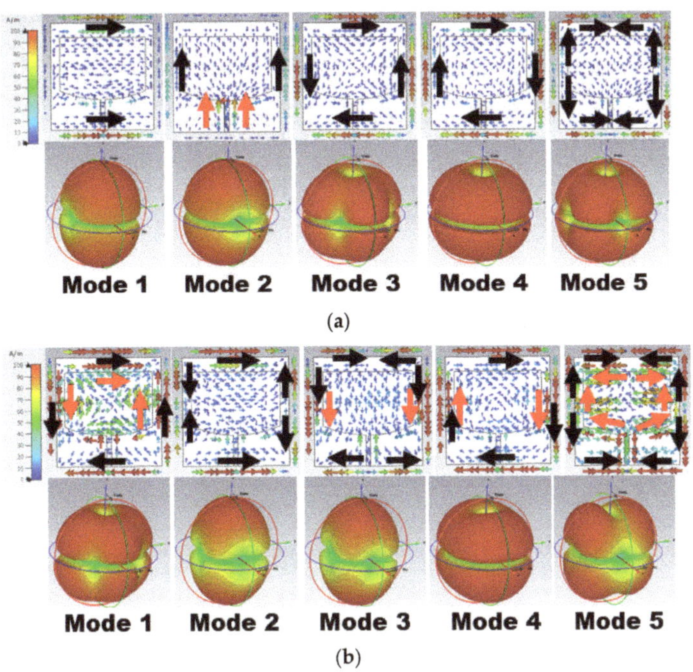

Figure 4. Eigenvector current J_n on the surface of Stage 2 and its respective radiation pattern at (**a**) 25.50 GHz and (**b**) 38 GHz.

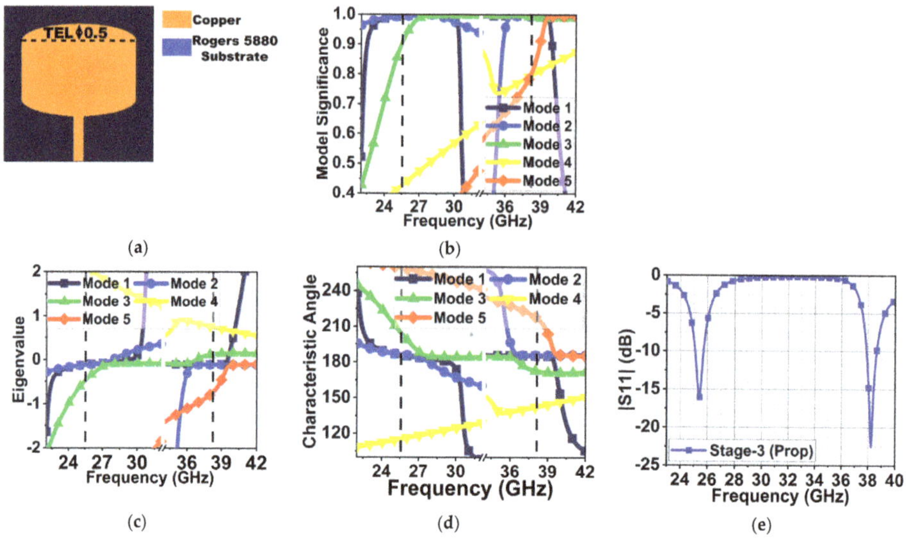

Figure 5. CMA of Mode 2–5 and reflection coefficients of the Stage-3 MS antenna. (**a**) Stage-1 antenna (dimensions in mm), (**b**) modal significance, (**c**) eigenvalues, (**d**) characteristic angle, and (**e**) reflection coefficient with port excitation.

The eigenvector current of Mode 1 and 2 at 25.5 GHz has an in-phase current resulting in bidirectional radiation, as shown in Figure 6a. At 38 GHz (Figure 6b), Mode 1 has an in-phase current along the vertical edges and an anti-phase current along the horizontal edges, resulting in bidirectional quad radiation. Similarly, in Mode 2, there are in-phase currents along horizontal edges and anti-phase currents along vertical edges. Though Mode 5 is largely capacitive, its surface current impacts the radiation patterns.

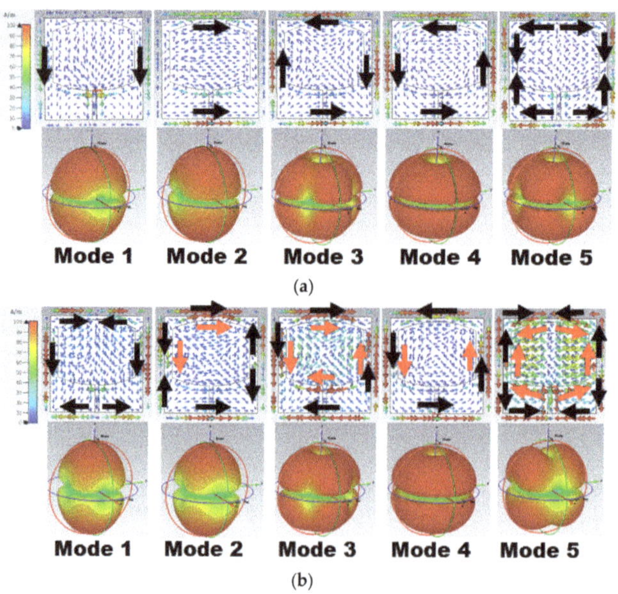

Figure 6. Eigenvector current J_n on the surface of Stage 3 and its respective radiation pattern at (**a**) 25.50 GHz and (**b**) 38 GHz.

3. Current Distribution with Port Excitation

From the above CMA and port excitation, it is shown that the structure generates dual resonance at 25.5 and 38 GHz. Figure 7 indicates the current oscillation at the respective frequencies when the port is excited. At 25.5 GHz, the current oscillation is along the $\pm x$-axis (i.e., vertical oscillation) because the patch length at the center is close to half-wavelength at 25.5 GHz, which is $0.4\lambda_0$ in this case. The current is at its maximum in the feed and the center of the patch along the x-axis. This gives rise to the TM_{10} mode at this frequency. However, at 38 GHz, the current oscillation in Figure 7b can be seen along the $\pm y$-axis because the patch width is close to half-wavelength at 38 GHz, which is $0.6\lambda_0$ in this case. Two current maximums along the patch width give rise to the TM_{02} mode at this frequency.

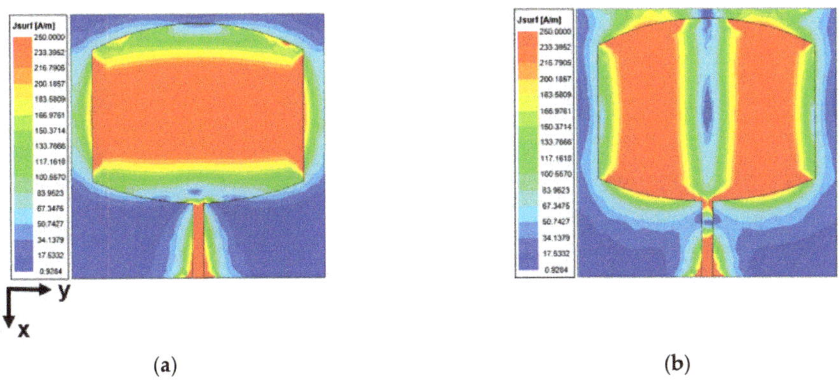

Figure 7. Current distribution with port excitation at (**a**) 25.5 GHz and (**b**) 38 GHz.

4. Parametric Analysis of Antenna Elliptical Structure

In this section, parametric analysis of two variables, BEL and TEL, is performed to study the effect of resonance change due to the radiator length change and its impact on the antenna impedance. Here, one variable is maintained at 0.5 mm, and the other varies from 0.1 to 0.6 mm. In the first case, TEL is kept at 0.5 mm, and BEL is varied. With the BEL at 0.1 mm, the overall length of the patch is reduced. Thus, the first resonance occurs at 27 GHz with an impedance matching of 22 dB because the impedance is purely real with 50 Ω at 27 GHz, as shown in Figure 8b. However, the reduced length impacts the impedance at 38 GHz with a poor impedance matching of 11 dB, as shown in Figure 8a. This is because the impedance at this frequency has a small inductance effect with a total impedance Z of 42 + j37 Ω. Further, with the patch length increase and a BEL from 0.2 to 0.6 mm, the first resonance drifts to lower frequencies with little decrease in impedance matching. This is because, as the patch length increases, the resonance decreases. At the second resonance, there is no effect on frequency drift. However, impedance matching is improved to a minimum of 30 dB.

In the second case, BEL is maintained at 0.5 mm, and TEL varies from 0.1 to 0.6 mm. A similar response can be observed compared to that of BEL. However, for TEL = 0.1 mm, the impedance matching at 25.5 and 38 GHz is good with 20 and 15 dB, as shown in Figure 9a. At 25.5 and 38 GHz, the impedance is real and close to 50 Ω, as displayed in Figure 9b,c. With a further increase in TEL from 0.2 to 0.6 mm, the first resonance is decreasing, and the second resonance impedance is improving, similar to the earlier case.

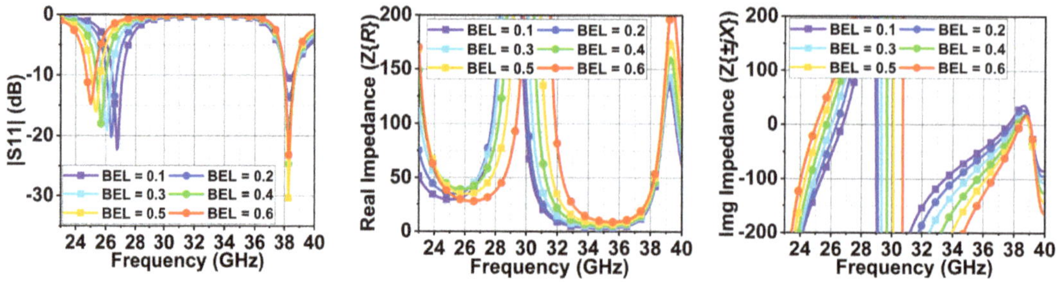

(a) (b) (c)

Figure 8. Variation in (**a**) reflection coefficient |S11| due to a change in the variable BEL, (**b**) Real and (**c**) imaginary impedance values at respective BEL values.

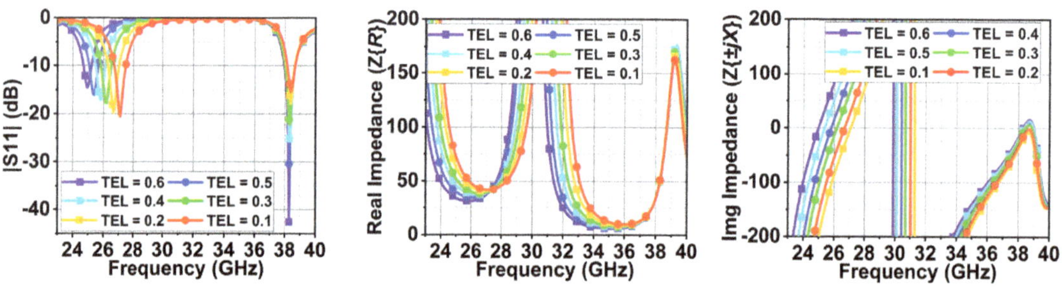

(a) (b) (c)

Figure 9. Effect of varying TEL on (**a**) the reflection coefficient $|S_{11}$, (**b**) the real part of the impedance, and (**c**) the imaginary parts of impedance.

5. Results and Discussion

5.1. Reflection Coefficients and Radiation Patterns in Free Space

The proposed antenna was fabricated, as shown in Figure 10a. The antenna reflection coefficient $|S_{11}|$ was measured using the Keysight vector network analyzer, which ranges from 300 KHz to 44 GHz. The connectors used also supported RF signal feeding up to 50 GHz. The $|S_{11}|$ measurement setup is shown in Figure 10b. The CMA results indicate that Mode 1 and 2 are resonant, with Mode 3 partially contributing to the resonance at 25.5 GHz with a bandwidth of 24–27 GHz. At 38 GHz, Mode 1–3 are resonant with a bandwidth of 26.2–39.3 GHz. The high-frequency simulation software (HFSS) version 21.0 results illustrate the occurrence of two resonances at 25.5 and 38 GHz with a bandwidth of 750 MHz (25–25.75 GHz) and 1 GHz (37.92–38.92 GHz). The measured |S11| shows resonance at 25.85 and 38 GHz with bandwidths of 1.5 GHz (25–26.5 GHz) and 2.5 GHz (37–39.5 GHz), as displayed in Figure 11a. The deviation in the measured results may be due to fabrication tolerance, cable, and connector losses. The procured connectors (the best available in the market) also have a 1.2 mm gap between the center signal and ground lead. When the antenna is fitted inside, a gap of 0.94 mm still exists that is filled with excessive soldering lead, as shown in Figure 12. Above all, the proposed antenna is very small and lacks industrial standard soldering tools, necessitating manual soldering that causes a lead spread on the ground plane. All of this impacts the antenna's performance, causing deviations in the measurement results, as shown in Figure 11. However, the measured results are considerable compared to the simulated results. The antenna has an almost constant simulated and measured gain of 7.4 and 7.9 dBi at first resonance. Nonetheless, at second resonance, the gain almost linearly increases with the maximum simulated and measured gains of 7.4 and 7.3 dBi, as shown in Figure 11b.

(a) (b)

Figure 10. Photos of the antenna and the reflection coefficient measurements. (**a**) Photos of the antenna and (**b**) the measurement setup of the |S11| parameter.

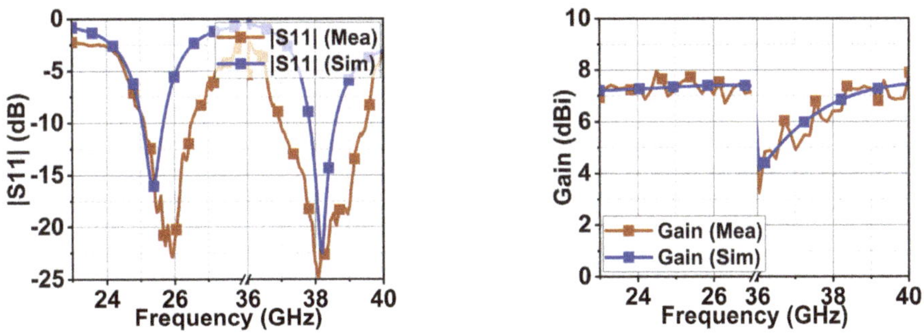

(a) (b)

Figure 11. Simulated and measured reflection coefficients and gain. (**a**) $|S_{11}|$ and (**b**) gain.

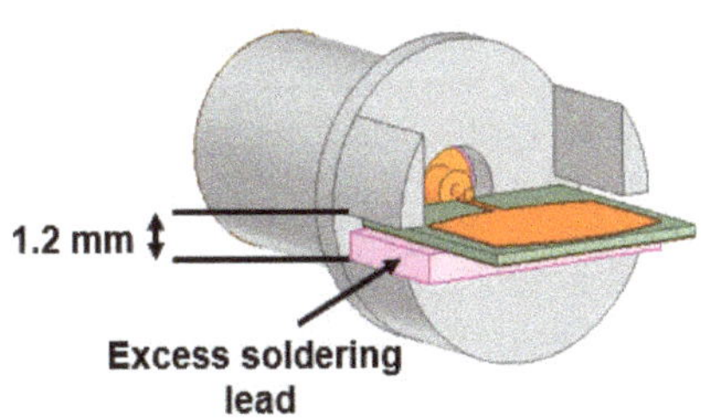

Figure 12. 3D representation of the proposed antenna with the SMA connector.

As discussed in the current distribution section, the proposed antenna generates TM_{10} and TM_{02} at 25.5 and 38 GHz, respectively. Due to these modes, the radiation is a single beam in the broadside direction at 25.5 GHz, whereas two beams at 38 GHz are tilted at 36°. Therefore, the radiation pattern is measured on the XZ- and YZ-planes at 25.5 GHz (Figure 13a,b), whereas, at 38 GHz, only the YZ-plane is considered (Figure 13c). The simulated and measured half-power-beamwidths (HPBWs) at 25.5 GHz are 78° and 62° on the XZ-plane. On the YZ-plane, they are 82° and 62°. At 38 GHz, the simulated and measured HPBWs are 58° and 45°.

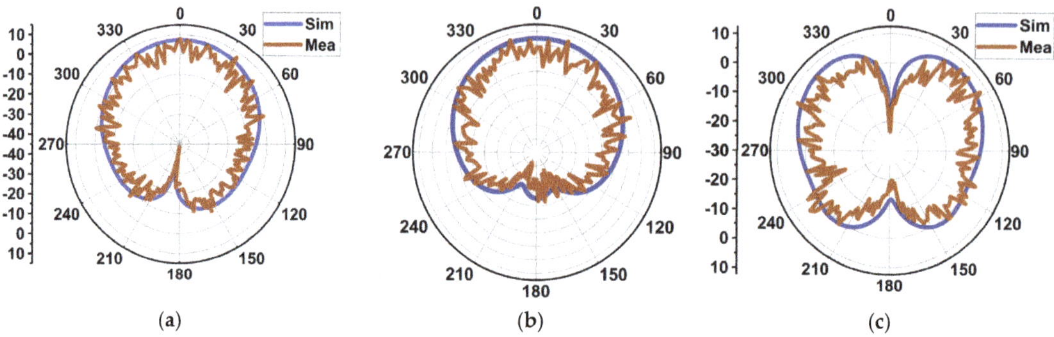

Figure 13. Simulated and measured radiation patterns. (**a**) The XZ-plane at 25.5 GHz, (**b**) the YZ-plane at 25.5 GHz, and (**c**) the YZ-plane at 38 GHz.

5.2. Antenna Performance with Watch Case

Since the proposed antenna is for SW applications, its performance was studied by placing it inside a round dial watch case of 23 mm radius and a height of 9 mm, as displayed in Figure 14. The watch body case was considered as stainless steel with a relative permittivity of 1. The watch top was covered with reinforced glass of PTFE material with a relative permittivity of 2.5. The antenna was placed at a height of 2.5 mm from the bottom of the watch case. At 25.5 GHz, a slight decrease in impedance occurs due to the formation of a negative image current on the metallic surface of the watch case [21,22]. However, at 38 GHz, the impedance is significantly increased. The reflection coefficient $|S_{11}|$ is shown in Figure 15. Nonetheless, other than an increase in the front-to-back ratio of radiation due to the metallic casing, the antenna performance is not strongly affected. The electric field distribution in and around the watch case is presented in Figures 16 and 17. On the XY-plane at 25.5 GHz, the E-field is concentrated inside the watch case mostly along the x-axis due to TM_{10}, as illustrated in Figure 16a. Outside the case, a minor E-field distribution can be seen at the front and back of the watch lug. Figure 16b,c show the E-field along the XZ- and YZ-planes. It is shown that the E-field intensity is in the broadside direction. The radiation pattern in Figure 16d shows a directional beam with an HPBW of 30° on the XZ- and YZ-planes with a maximum gain of 9.9 dBi. In the case of 38 GHz, the E-field distribution is on the y-axis of the XY-plane due to TM_{02}, as depicted in Figure 17a. The E-field is mostly concentrated inside the case, and some intensity can be seen along and opposite the watch crown side. Figure 17b,c show the E-field on the XZ- and YZ-planes. The intensity of the field is maximum on the YZ-plane in two directions, approximately ±36°. The radiation pattern with two beams on the YZ-plane is shown in Figure 17d with an HPBW of 46° and a maximum gain of 10.9 dBi.

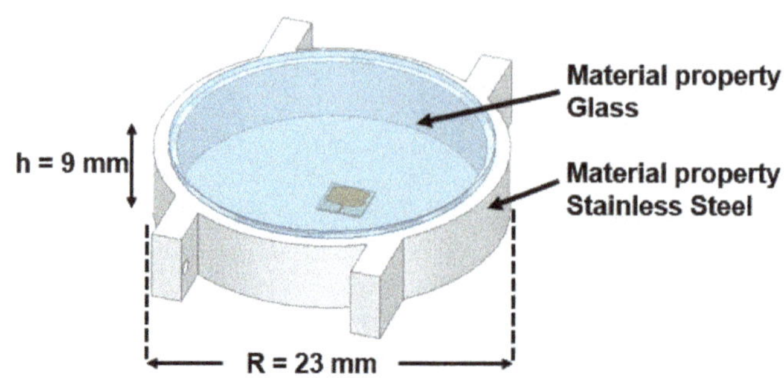

Figure 14. Model of the smartwatch with the antenna inside.

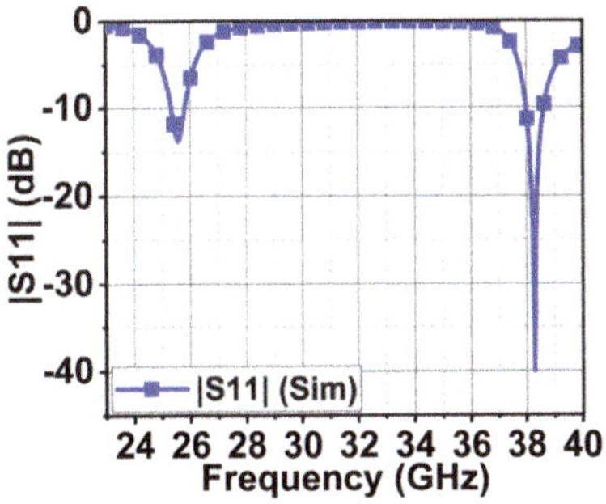

Figure 15. Reflection coefficient $|S_{11}|$ antenna response when placed inside the watch case.

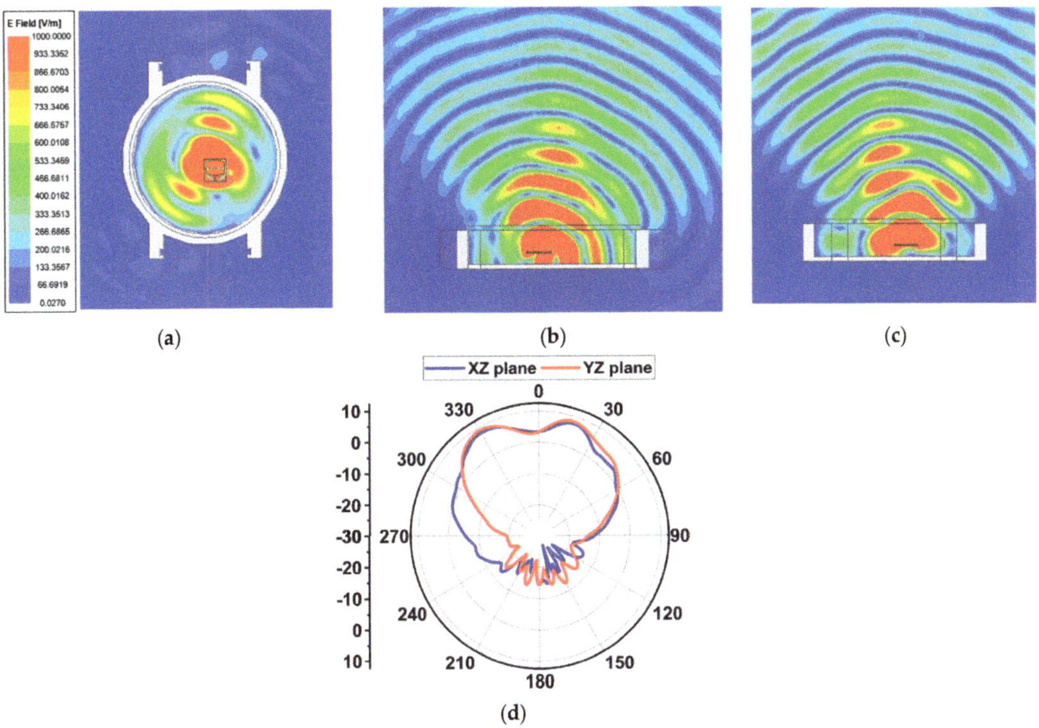

Figure 16. Electric field distribution and radiation patterns at 25.5 GHz with the antenna inside the watch case. (**a**) The XY-plane, (**b**) the XZ-plane, (**c**) the YZ-plane, and (**d**) the radiation pattern on the YZ-plane.

5.3. SAR Analysis

This section presents the study of the RF energy absorption on human tissue generated from the designed antenna, as the proposed design is for wearable smartwatch applications. As per new guidelines of the International Commission for Non-Ionizing Radiation Protection (ICNIRP), the absorption rate is mainly at the skin layer with less than 1 cm for frequencies >6 GHz [23,24]. However, a three-layer phantom model was developed with skin, fat, and muscle to study the absorption rate, as displayed in Figure 18. The dielectric properties of human tissue vary over the body and the different frequencies. Therefore, the tissue parameters are considered from a verified dataset of the IT'IS foundation [25]. Table 1 presents the dielectric property, electrical conductivity, and density of tissues considered for our analysis at 25.5 and 38 GHz. Due to the limitation of computation resources, the watch case is replaced by the watch back case only, which is of a 1 mm thickness and a 23 mm radius. The antenna is placed 2.5 mm above this case. The analysis shows that the SAR is 0.063 and 0.0206 W/kg at 25.5 and 38 GHz, respectively, as displayed in Figure 19. There are two reasons for the low SAR value: (i) The antenna has a full ground plane, and (ii) the antenna is backed by a watch back case that behaves as a reflector for electromagnetic energy. Consequently, the results are satisfactory, suggesting that the antenna is suitable for wearable smartwatch applications.

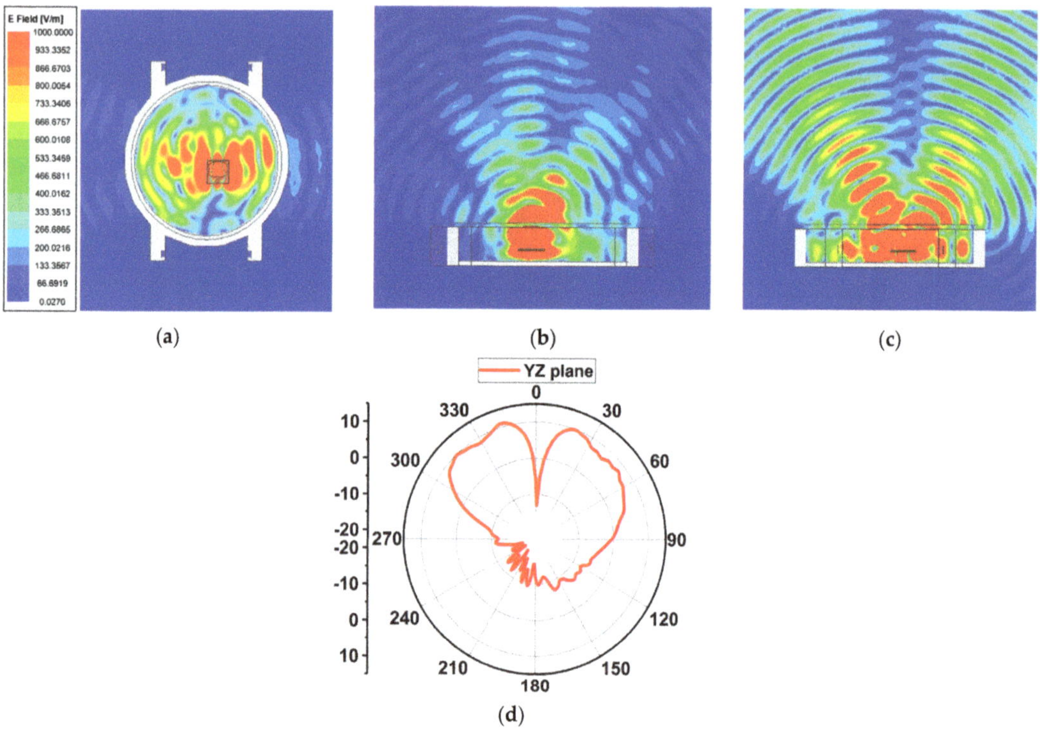

Figure 17. Electric field distribution and radiation patterns at 38 GHz with the antenna inside the watch case. (**a**) The XY-plane, (**b**) the XZ-plane, (**c**) the YZ-plane, and (**d**) the radiation pattern on the YZ-plane.

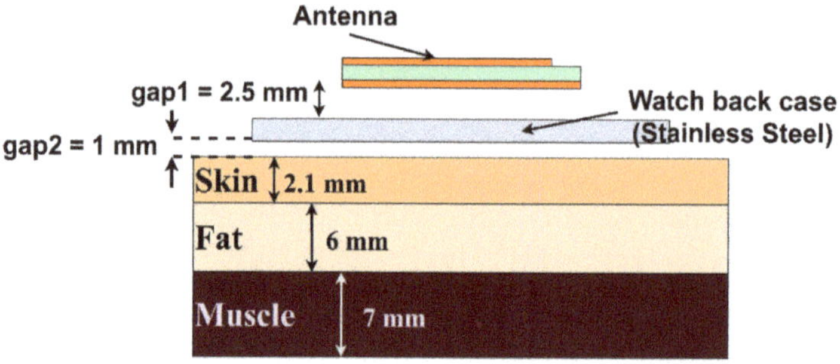

Figure 18. Phantom model of human tissue to analyze the RF energy penetration.

Table 1. Human tissue properties at different frequencies.

		25.5 GHz		38 GHz	
Tissue	**Density (kg/m^3)**	**Permittivity**	**Electrical Conductivity (s/m)**	**Permittivity**	**Electrical Conductivity (s/m)**
Skin	1109	18	24	12.3	31
Fat	911	6.34	4.66	5.33	6.36
Muscle	1090	26.2	31.1	19.1	41.8

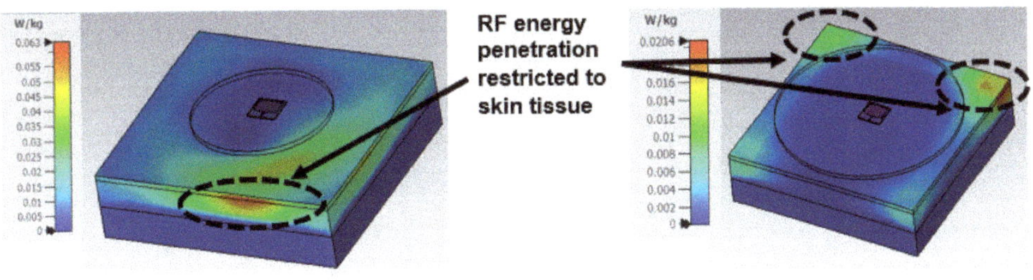

Figure 19. SAR analysis results (**a**) at 25.5 GHz and (**b**) at 38 GHz.

5.4. Link Budget Analysis

The smartwatch on the human wrist experiences complex three-dimensional movement during various day-to-day activities. Therefore, error-free transmission and reception are essential for seamless, smooth communication between smartwatches and mobile/IoT devices during these movements. A link budget was studied between the mobile and proposed smartwatch antenna under a line-of-sight (LOS) environment at various rotational degrees. A standard dipole antenna was considered as a reference antenna (transmitting antenna) inside the mobile case with a gain (G_{td}) of 1 dBi. The receiving antenna is the proposed antenna (antenna under test (AUT)). The receiver antenna gain (G_{tr}) is 9.9 and 10.9 dBi at 25.5 and 38 GHz (placed inside the watch case, as discussed in the previous section). In our case, the reference antenna and AUT were separated by a 50 cm distance. The AUT was stationary but rotated over the x-axis from 0° to ±180°, as depicted in Figure 20. The received signal at various rotational angles was computed using the shooting and bouncing ray technique in HFSS Savant, considering the losses such as the path loss exponent effect and the shadowing effect (LS) [26]. Therefore, Equation (2) [26,27] can calculate the received power as

$$P_{rw} = P_{td} + G_{td} + G_{tr} - LS \tag{2}$$

In the analysis, the power of the transmitting antenna (P_{td}) considered is 30 mW. The required power (R_{Pw}) to estimate the link budget for an ideal binary-phase-shift-keying (BPSK) with an energy-to-noise ratio (E_b/N_0) of 9.6 dB is given by Equation (3):

$$R_{Pw}\,(dB) = \frac{E_b}{N_0}\,(dB) + KT + B_r\,(dB) \tag{3}$$

where K is the Boltzmann constant with a value of 1.38×10^{-23}, T is the temperature in Kelvin (in our case, 290 K), and B_r represents the various bit rates supported for reliable communication. The link margin is the difference between the actual received power P_{rw} and required power R_{Pw} given by Equation (4) [27,28]:

$$LM\,(dB) = P_{rw}\,(dB) - R_{Pw}\,(dB) \tag{4}$$

Assuming that a link margin of 20 dB is sufficient for error-free communication, the proposed antenna with the highest data rate of 1 Gbps is supported with a link margin of >20 from 0° to −180° angle orientation, as depicted in Figure 21a,b. At these angles, both the reference and AUT antenna have good LOS conditions; in fact, direct LOS occurs at −90°. From 0° to 180°, specifically from 45° to 135°, the AUT antenna experiences a shadowing effect due to the metallic case of the watch. As a result, the link margin significantly deteriorates, as shown in Figure 21a,b. Nonetheless, the performance of the proposed antenna is satisfactory, with a data rate support of 1 Gbps.

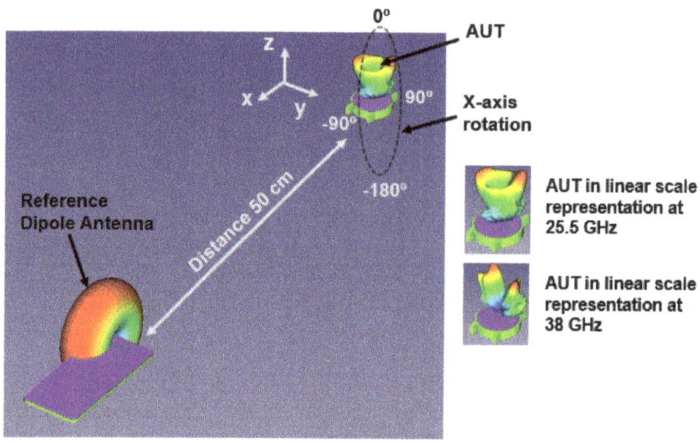

Figure 20. Link budget estimation environment under line-of-sight conditions.

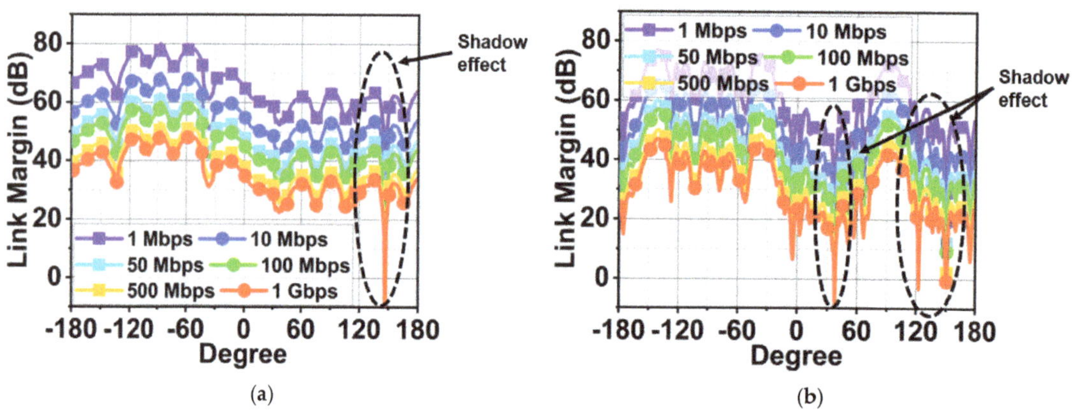

Figure 21. Estimated link budget between the ideal dipole antenna and the proposed antenna at 50 cm distance for various rotational angles for the two bands. (**a**) 25.5 GHz and (**b**) 38 GHz.

5.5. Comparative Analysis

The proposed antenna performance was compared with the existing designs in Table 2. The proposed antenna is relatively compact compared to most of the other antennas in the table. The antenna also has a comparatively optimal gain. The proposed antenna with the bottom watch case has the lowest SAR value. Most of the antennas in the table are planar antennas; however, due to partial ground/defected ground structure (to improve the bandwidth), the antenna structure replicates monopole behavior. As a result, these are labeled as monopole antennas. In our case, the antenna has a full ground plane depicting a perfect patch antenna. The proposed antenna's link budget was also analyzed, and it was demonstrated that the antenna can deliver 1 Gbps of data at a 20 dB link margin, which other designs have not performed.

Table 2. Comparative analysis of proposed antenna with existing designs.

Ref	Ant Type	Dim (mm^2)	Res (GHz)	BW (GHz)	Gain (dBi)	SA (W/kg)	Radiation	LM Analysis
[11]	Mono	$0.37\lambda_0 \times 27\lambda_0$	27/38/60	22–30/38–39.5/56–61	5.29/7.49/9	0.9/0.62/0.74	Bi-Dir	No
[29]	Mono	$1.1\lambda_0 \times 0.86\lambda_0$	28.5/38	26–40.5	3.8	NA	Bi-Dir	No
[30]	Mono	$2.41\lambda_0 \times 2\lambda_0$	60	58.2–60.5	9.9	NA	Multi-Beam	No
[31]	Mono	$1.36\lambda_0 \times 1.22\lambda_0$	28/38	22–29/37–40	5	NA	Omni	No
[32]	Mono	$0.75\lambda_0 \times 1.24\lambda_0$	37	36.7–37.4	NA	NA	Broadside	No
[33]	Mono	$1.12\lambda_0 \times 1.12\lambda_0$	28/40	24–32/38–41	4.5	NA	Bi-Dir	No
Proposed	Patch	$0.51\lambda_0 \times 0.51\lambda_0$	25.5/38	25–26.5/37–39.5	7.4/7.9	0.063/0.0206	Broadside	Yes

Note: Ref: reference; Ant: antenna; Dim: dimension; Res: resonance; LM: link margin; BW: bandwidth.

6. Future Scope

In line with the presented research work, other contributions can be made: (1) The RF energy exposure by the proposed antenna with SWs can be studied in real time by mimicking the phantom model. (2) In the current work, the effect of the SW screen was neglected. In future work, the effect of the SW screen, which is supported by PCB and has PEC properties, can be considered in the analysis. (3) Further, link margin analysis can be performed by considering the human body between SWs and mobile phones, since at the mmWave level, the human body creates obstruction, leading to NLOS communication.

7. Conclusions

This article presents a compact dual-band antenna operating at 25.5 and 38 GHz bands. Due to its compact design, it is suitable for smartwatch and IoT applications. The antenna structure has generated two modes at 25.5 GHz and three modes at 38 GHz, Mode 1 and 2 and Mode 1–3, which are analyzed using characteristic mode theory. Apart from these modes, the antenna generated traverse TM_{10} and TM_{02} modes at the respective bands. The antenna has a good measured bandwidth of 1.5 and 2.5 GHz with gains of 7.4 and 7.9 dBi. Due to traverse modes, the antenna has radiation in the broadside with single and dual beams at their respective bands. The antenna showed a sustainable $|S_{11}|$ level when tested with the watch case. This also improved the gain to 9.9 and 10.9 dBi. The results of SAR analysis indicate minimal RF energy penetration to human tissue, which makes it suitable for wearable applications. The link budget estimation demonstrated the highest data rate transfer of 1 Gbps for a link margin of 20 dB.

Author Contributions: Conceptualization, P.S.B.G., P.R.M. and T.A.; methodology, P.S.B.G.; software, P.S.B.G. and S.P.; validation, S.T.S., V.K.P. and V.K.J.; writing—original draft preparation, P.S.B.G.; writing—and editing, T.A. and P.R.M.; supervision, S.T.S. All authors have read and agreed to the published version of the manuscript.

Funding: This research received no external funding.

Institutional Review Board Statement: Not Applicable.

Informed Consent Statement: Not Applicable.

Data Availability Statement: All the data are contained within the manuscript.

Conflicts of Interest: The authors declare no conflict of interest.

References

1. Chandel, R.S.; Sharma, S.; Kaur, S.; Singh, S.; Kumar, R. Smart Watches: A Review of Evolution in Bio-Medical Sector. *Mate Today Proc.* **2022**, *50*, 1053–1066. [CrossRef]
2. Global: Smartwatches Number of Users 2018–2027. Available online: https://www.statista.com/forecasts/1314339/worldwide users-of-smartwatches (accessed on 11 November 2023).
3. Hao, T.; Bi, C.; Xing, G.; Chan, R.; Tu, L. MindfulWatch: A Smartwatch-Based System for Real-Time Respiration Monitoring During Meditation. *Proc. ACM Interact. Mob. Wearable Ubiquitous Technol.* **2017**, *1*, 1–19. [CrossRef]
4. Xintarakou, A.; Sousonis, V.; Asvestas, D.; Vardas, P.E.; Tzeis, S. Remote Cardiac Rhythm Monitoring in the Era of Smar Wearables: Present Assets and Future Perspectives. *Front. Cardiovasc. Med.* **2022**, *9*, 853614. [CrossRef] [PubMed]
5. Kwon, J.; Jo, Y.-Y.; Lee, S.Y.; Kang, S.; Lim, S.-Y.; Lee, M.S.; Kim, K.-H. Artificial Intelligence-Enhanced Smartwatch ECG for Hear Failure-Reduced Ejection Fraction Detection by Generating 12-Lead ECG. *Diagnostics* **2022**, *12*, 654. [CrossRef] [PubMed]
6. He, J.; Ou, J.; He, A.; Shu, L.; Liu, T.; Qu, R.; Xu, X.; Chen, Z.; Yan, Y. A New Approach for Daily Life Blood-Pressure Estimatior Using Smart Watch. *Biomed. Signal Process. Control* **2022**, *75*, 103616. [CrossRef]
7. Xu, Z.; Wang, Y. Design of Dual-Band Antenna for Metal-Bezel Smartwatches with Circular Polarization in GPS Band and Low Wrist Effect. *IEEE Trans. Antennas Propagat.* **2023**, *71*, 4651–4662. [CrossRef]
8. Liao, C.-T.; Yang, Z.-K.; Chen, H.-M. Multiple Integrated Antennas for Wearable Fifth-Generation Communication and Internet of Things Applications. *IEEE Access* **2021**, *9*, 120328–120346. [CrossRef]
9. Xiao, B.; Wong, H.; Wu, D.; Yeung, K.L. Design of Small Multiband Full-Screen Smartwatch Antenna for IoT Applications. *IEEE Internet Things J.* **2021**, *8*, 17724–17733. [CrossRef]
10. Mallat, N.K.; Ishtiaq, M.; Ur Rehman, A.; Iqbal, A. Millimeter-Wave in the Face of 5G Communication Potential Applications. *IETE J. Res.* **2022**, *68*, 2522–2530. [CrossRef]
11. Ahmad, S.; Boubakar, H.; Naseer, S.; Alim, M.E.; Sheikh, Y.A.; Ghaffar, A.; Al-Gburi, A.J.A.; Parchin, N.O. Design of a Tri-Band Wearable Antenna for Millimeter-Wave 5G Applications. *Sensors* **2022**, *22*, 8012. [CrossRef]
12. Ruchi; Patnaik, A.; Kartikeyan, M.V. Compact Dual and Triple Band Antennas for 5G-IOT Applications. *Int. J. Microw. Wirel Technol.* **2022**, *14*, 115–122. [CrossRef]
13. Garbacz, R.; Turpin, R. A Generalized Expansion for Radiated and Scattered Fields. *IEEE Trans. Antennas Propagat.* **1971**, *19* 348–358. [CrossRef]
14. Yee, A.; Garbacz, R. Self-and Mutual-Admittances of Wire Antennas in Terms of Characteristic Modes. *IEEE Trans. Antennas Propagat.* **1973**, *21*, 868–871. [CrossRef]
15. Kim, G.; Kim, S. Design and Analysis of Dual Polarized Broadband Microstrip Patch Antenna for 5G mmWave Antenna Module on FR4 Substrate. *IEEE Access* **2021**, *9*, 64306–64316. [CrossRef]
16. Adams, J.J.; Genovesi, S.; Yang, B.; Antonino-Daviu, E. Antenna Element Design Using Characteristic Mode Analysis: Insights and Research Directions. *IEEE Antennas Propag. Mag.* **2022**, *64*, 32–40. [CrossRef]
17. Gao, G.; Zhang, R.-F.; Geng, W.-F.; Meng, H.-J.; Hu, B. Characteristic Mode Analysis of a Nonuniform Metasurface Antenna for Wearable Applications. *Antennas Wirel. Propag. Lett.* **2020**, *19*, 1355–1359. [CrossRef]
18. Li, H.; Tan, Y.; Lau, B.K.; Ying, Z.; He, S. Characteristic Mode Based Tradeoff Analysis of Antenna-Chassis Interactions for Multiple Antenna Terminals. *IEEE Trans. Antennas Propagat.* **2012**, *60*, 490–502. [CrossRef]
19. Mohanty, A.; Behera, B.R. Characteristics Mode Analysis: A Review of Its Concepts, Recent Trends, State-Of-The-Art Developments and Its Interpretation with A Fractal UWB MIMO Antenna. *Prog. Electromagn. Res. B* **2021**, *92*, 19–45. [CrossRef]
20. Balanis, C.A. *Antenna Theory: Analysis and Design*, 4th ed.; John Wiley & Sons: Hoboken, NJ, USA, 2016; ISBN 978-1-118-64206-1.
21. Ukkonen, L.; Sydanheimo, L.; Kivikoski, M. Effects of Metallic Plate Size on the Performance of Microstrip Patch-Type Tag Antennas for Passive RFID. *Antennas Wirel. Propag. Lett.* **2005**, *4*, 410–413. [CrossRef]
22. Klionovski, K.; Shamim, A. Back Radiation Suppression Through a Semitransparent Ground Plane for a Millimeter-Wave Patch Antenna. *IEEE Trans. Antennas Propagat.* **2017**, *65*, 3935–3941. [CrossRef]
23. Redmayne, M.; Maisch, D.R. ICNIRP Guidelines' Exposure Assessment Method for 5G Millimetre Wave Radiation May Trigger Adverse Effects. *Int. J. Environ. Res. Public Health* **2023**, *20*, 5267. [CrossRef] [PubMed]
24. Taguchi, K.; Kodera, S.; Hirata, A.; Kashiwa, T. Computation of Absorbed Power Densities in High-Resolution Head Models by Considering Skin Thickness in Quasi-Millimeter and Millimeter Wave Bands. *IEEE J. Electromagn. RF Microw. Med. Biol.* **2022**, *6*, 516–523. [CrossRef]
25. IT'IS Foundation. TISSUE DB. Database at a Glance. Available online: https://itis.swiss/virtual-population/tissue-properties/database/ (accessed on 4 April 2023).
26. Bommisetty, L.; Pawar, S.; Venkatesh, T.G. Performance Analysis of Random Access Mechanism in 5G Millimeter Wave Networks: Effect of Blockage, Shadowing and Mobility. *IEEE Access* **2022**, *10*, 69091–69105. [CrossRef]
27. Shariff, B.G.P.; Naik, A.A.; Ali, T.; Mane, P.R.; David, R.M.; Pathan, S.; Anguera, J. High-Isolation Wide-Band Four-Element MIMO Antenna Covering Ka-Band for 5G Wireless Applications. *IEEE Access* **2023**, *11*, 123030–123046. [CrossRef]
28. Iqbal, A.; Al-Hasan, M.; Mabrouk, I.B.; Nedil, M. A Compact Implantable MIMO Antenna for High-Data-Rate Biotelemetry Applications. *IEEE Trans. Antennas Propagat.* **2022**, *70*, 631–640. [CrossRef]

29. Munir, M.E.; Al Harbi, A.G.; Kiani, S.H.; Marey, M.; Parchin, N.O.; Khan, J.; Mostafa, H.; Iqbal, J.; Khan, M.A.; See, C.H.; et al. A New Mm-Wave Antenna Array with Wideband Characteristics for Next Generation Communication Systems. *Electronics* **2022**, *11*, 1560. [CrossRef]
30. Mneesy, T.S.; Hamad, R.K.; Zaki, A.I.; Ali, W.A.E. A Novel High Gain Monopole Antenna Array for 60 GHz Millimeter-Wave Communications. *Appl. Sci.* **2020**, *10*, 4546. [CrossRef]
31. Ali Esmail, B.; Koziel, S. High Isolation Metamaterial-Based Dual-Band MIMO Antenna for 5G Millimeter-Wave Applications. *AEU Int. J. Electron. Commun.* **2023**, *158*, 154470. [CrossRef]
32. Khan, J.; Ullah, S.; Ali, U.; Tahir, F.A.; Peter, I.; Matekovits, L. Design of a Millimeter-Wave MIMO Antenna Array for 5G Communication Terminals. *Sensors* **2022**, *22*, 2768. [CrossRef]
33. Munir, M.E.; Kiani, S.H.; Savci, H.S.; Marey, M.; Khan, J.; Mostafa, H.; Parchin, N.O. A Four Element Mm-Wave MIMO Antenna System with Wide-Band and High Isolation Characteristics for 5G Applications. *Micromachines* **2023**, *14*, 776. [CrossRef]

sensors MDPI

Article

Wide-Angle Beam Steering Closed-Form Pillbox Antenna Fed by Substrate-Integrated Waveguide Horn for On-the-Move Satellite Communications

Muhammad Ikram [1,2], Kamel Sultan [1,*], Ahmed Toaha Mobashsher [1], Mahdi Moosazadeh [1] and Amin Abbosh [1]

[1] School of Electrical Engineering and Computer Science, The University of Queensland (UQ), Brisbane 4072, Australia; m.ikram@uq.net.au (M.I.); a.mobashsher@uq.edu.au (A.T.M.); m.moosazadeh@uq.edu.au (M.M.); a.abbosh@uq.edu.au (A.A.)

[2] College of Engineering and Applied Science (CEAS), American University of Kuwait, Safat 13034, Kuwait

* Correspondence: k.sultan@uq.edu.au

Abstract: Wide-angle mechanical beam steering for on-the-move satellite communications is presented in this paper based on a closed-form pillbox antenna system. It includes three main parts: a fixed-feed part, which is a substrate-integrated waveguide (SIW) horn with an extended aperture attached to a parabolic reflector; a novel quasi-optical system, which is a single coupling slot alongside and without spacing from the parabolic reflector; and a radiating disc, which is a leaky-wave metallic pattern. To make the antenna compact, pillbox-based feeding is implemented underneath the metallic patterns. The antenna is designed based on a substrate-guided grounded concept using leaky-wave metallic patterns operating at 20 GHz. Beam scanning is achieved using mechanical rotation of the leaky-wave metallic patterns. The proposed antenna has an overall size of 340 × 335 × 2 mm^3, a gain of 23.2 dBi, wide beam scanning range of 120°, from −60° to +60° in the azimuthal plane, and a low side lobe level of −17.8 dB at a maximum scan angle of 60°. The proposed antenna terminal is suitable for next-generation ubiquitous connectivity for households and small businesses in remote areas, ships, unmanned aerial vehicles, and disaster management.

Keywords: flat panel antenna; low earth orbit satellites; on-the-move communications

1. Introduction

Recently, there has been a rise in the demand for satellite services primarily driven by the necessity for worldwide broadcasting and the establishment of networks for sharing data information [1–3]. These networks include digital radio, television, and broadband internet services. To meet the requirements of those services, the integration of wireless communications technology with low earth orbit (LEO) satellite communications systems is introduced [3–6]. So, a low-cost and compact antenna is required to connect with LEO satellites. This antenna can be mounted on ambulances, unmanned aerial vehicles (UAVs), trains, vehicles, ships, robots for remote sensing, and homes to provide satellite communications on-the-move (SCOM) services (see Figure 1). While the traditional parabolic antenna is a high-performance solution, its large size makes it unsuitable for most of those applications. Therefore, there is a growing need for flat, compact, lightweight, and mechanically robust antennas that align with the specific requirements of these systems [7–11]. Additionally, the antennas' gain and beam coverage range play a significant role in overcoming path loss in high-frequency bands and enhancing resolution [12–19]. For SCOM systems, it is essential to have an antenna that can balance high gain and a wide coverage area. Additionally, the antenna must be capable of beam steering to establish real-time satellite-to-ground connections for SCOM [20]. This is where the beam steering capability of the antenna plays an essential role.

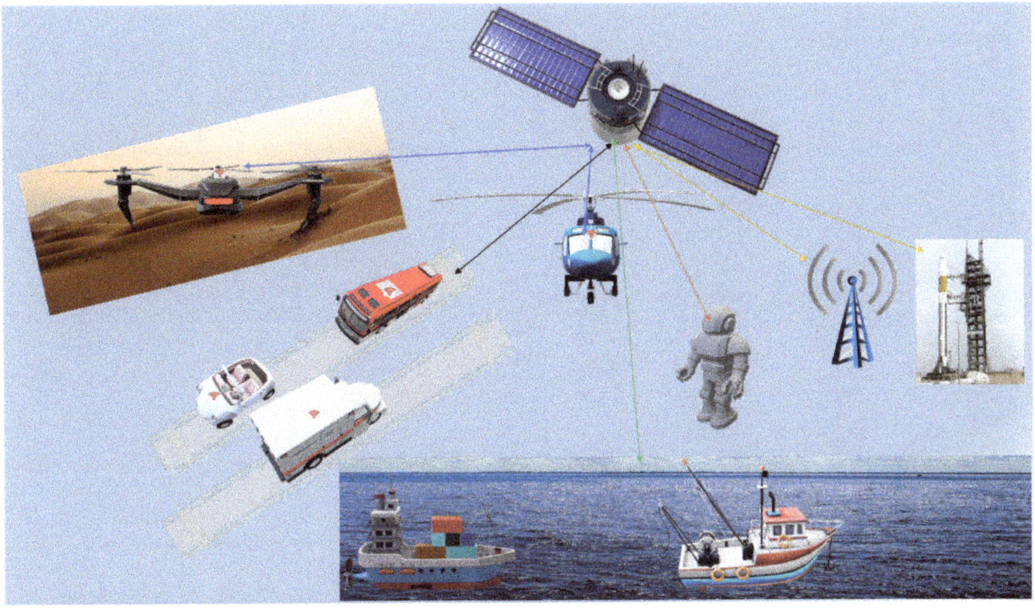

Figure 1. The conceptual scenario of satellite communications on-the-move services.

Beam-scanning antenna arrays can use either mechanical or electrical scanning approaches [21,22]. Electrical beam scanning can be achieved through switched antenna arrays or phased antenna arrays. Although these methods offer the advantages of low profile and rapid scanning speed, they are expensive, complex to implement, and suffer from high loss at higher frequencies, making them unsuitable for use in commercial terminal devices [23–26]. On the other hand, mechanical beam scanning is low-cost and offers high gain, but it is typically bulky and heavy and unsuitable for recent applications requiring compact designs [21,27].

Several different designs have been presented for beamforming feeding networks. These designs are based on passive technologies and do not rely on active components. Examples of these designs include the Rotman lens [28,29], Luneburg lens [12,30,31], Butler matrix [32,33], and pillbox reflector [34–36]. Pillbox reflector-based feeding designs can steer the beam across a wide bandwidth, but they require multiple feeding ports and provide limited beam coverage. The 3D-printed lens is unsuitable for applications requiring low-profile and conformal features as it has a high profile and fragile structure. In comparison, the Butler matrices achieve a low profile and wide scanning angle, making them a better option than lens antennas. However, Butler matrices face challenges when it comes to exciting a large number of antenna array elements [33,37,38].

In this paper, wide-angle beam steering is presented for the flat-panel antenna by a closed-form pillbox platform. To simplify the mechanical set-up and integrate appropriately with RF components for mechanical beam steering purposes, the antenna feed is fixed, while beam steering is achieved by rotating a metallic disc in front of the fixed feed. With a fixed feed, it avoids placing integrated horns in the focal plane of the parabolic reflector. A novel quasi-optical system is introduced to smoothly transition quasi-TEM mode propagation to the planar wavefront. First, to confine electromagnetic waves, an SIW-based horn with an extended aperture is attached to the parabolic reflector to create a closed-form platform. Second, a single coupling slot is created alongside and without spacing from the parabolic reflector to ensure an effective coupling and extend the bandwidth.

The remainder of this paper is organized as follows. Section 2 presents the overall design mechanism and discusses the individual part concept, while Section 3 provides and discusses the antenna performance in simulation and experimental environments and comparison with state-of-the-art designs. Finally, the paper is summarized in Section 4.

2. Flat Panel Antenna Design

The configuration of the flat panel antenna terminal is presented in Figure 2a, while the detailed structure is shown in Figure 2b,c. The antenna is stacked up with three double-sided Rogers *RO-4003C* substrates with a thickness (t) of 1.52 mm and a dielectric constant (ϵ_r) of 3.55 (see Figure 2b,c). To achieve SIW feeding, the bottom substrate (D_1) is in the shape of a rectangle and has two metallic layers on both sides (M_1 and M_2) to configure the parabolic structure of the pillbox. The middle substrate (D_2), on the other hand, has a ground plane for metallic patterns (M_2), as well as a coupling slot on the metallic structure (M_3) on the top of the substrate. The top substrate (D3) is shaped like a disc and contains metallic strip patterns that work as an antenna array (M4). In principle, the antenna has two main parts: (a) the radiating part (printed on the top layer (D3)), which consists of a rotating disc containing metallic patterns, and (b) the feeding part (a pillbox feeding system), which includes SIW feeding, a coupling slot (etched in the middle layer), and a parabolic reflector (etched through metallic vias between the bottom and middle layers). The disc has a radius of 150 mm, while the overall size of the antenna is $340 \times 335 \times 2$ mm^3. The working principle of the antenna is discussed in Section 2.1.

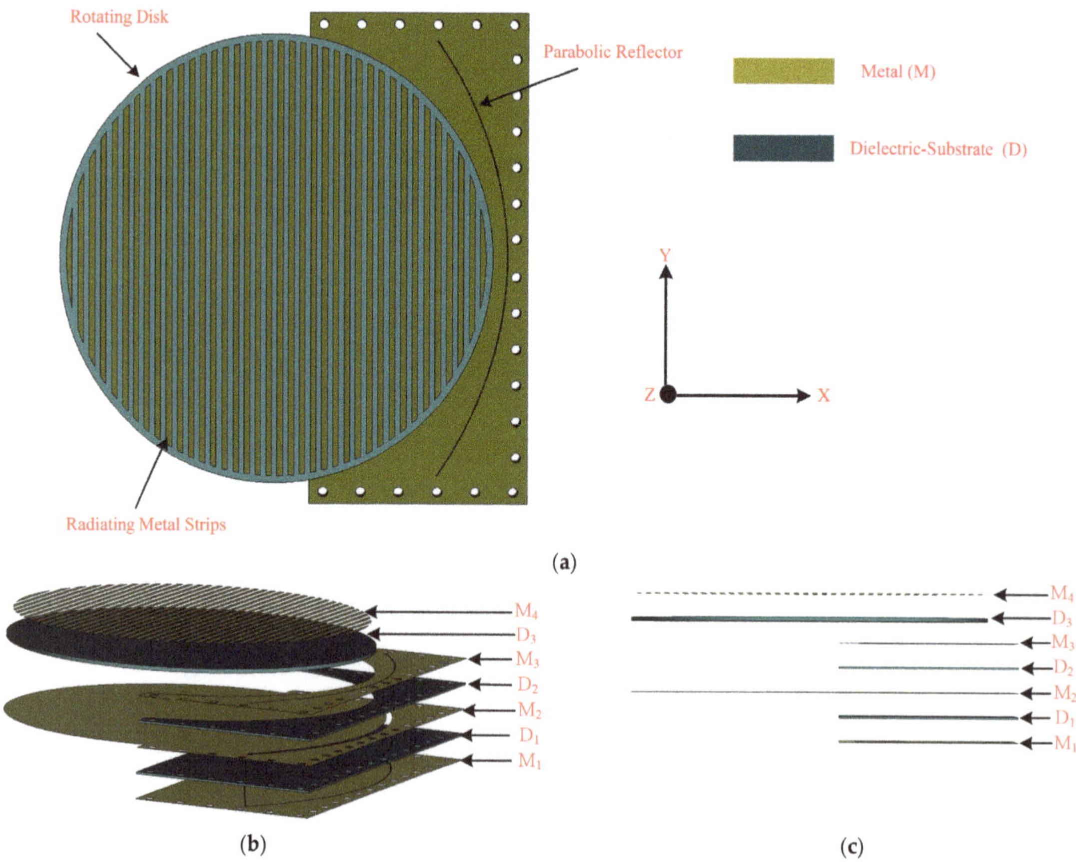

Figure 2. (**a**) The geometry of the proposed antenna and (**b**) its detailed configuration showing multiple layers and (**c**) side view of the layers, where M and D refer to metal and dielectric layers, respectively.

2.1. Design Principles

The antenna is designed based on a substrate-guided grounded (SGG) concept. This technique requires metallic patterns, feed, guiding substrate, and a ground plane (reflector), as depicted in Figure 3. As can be seen from Figure 3a, the metallic patterns are printed on the top side of the substrate, whereas the ground plane is located on the bottom side of the substrate. The metallic patterns are excited via the waveguide port in the simulator CST microwave studio (see Figure 3b). The waveguide port generates a plane wave, as shown in Figure 3b, which is guided through the substrate and hits the metallic patterns from the edge of the substrate. Once the metallic patterns are excited through the plane wave, it starts radiating the waves into the air, as shown in Figure 3c. This type of radiation is referred to as leaky waves. This structure provides highly directive radiation patterns with low-side lobes, as shown in Figure 3c. Beam scanning is accomplished through the utilization of the mechanical rotation of metallic patterns. Thus, the metallic patterns in the proposed design are printed in a circular shape. The final design can scan the beam from $-60°$ to $+60°$, which is an extraordinary performance compared to the existing literature [4,35,39,40]. The detailed results are presented in Section 3.

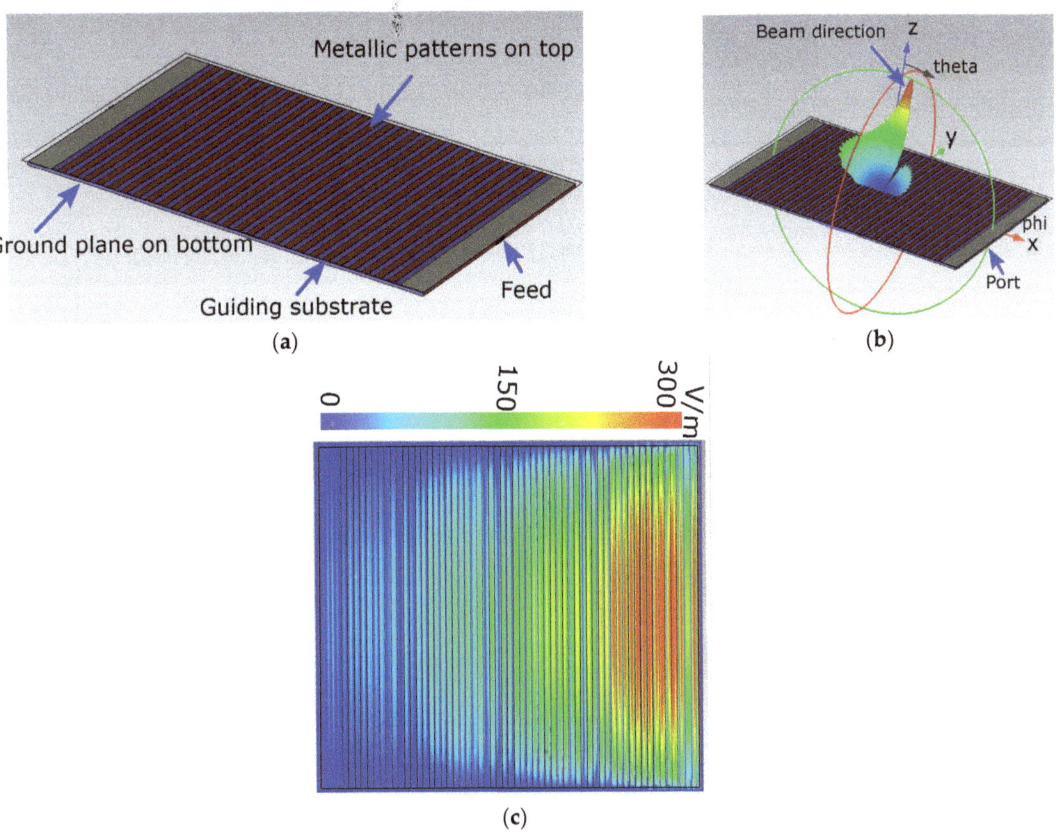

Figure 3. Antenna utilizing substrate-guided grounded concept. (**a**) Configuration of the antenna metallic patterns excited from waveguide port, (**b**) electric field distributions, and (**c**) simulated 3D highly directive radiation patterns obtained at 20 GHz.

2.2. Final Design

In order to develop the antenna based on the SGG method, a circular disc with a radius of 150 mm and a pillbox feeding (quasi-optical) system are developed. Figure 4 shows the details of each layer in the proposed design before assembling the antenna structure. The proposed antenna consists of four conducting layers (M_1:M_4) interleaved by three dielectric substrates (D1:D3). Table 1 shows the parameter of the antenna where all the dimensions are in (mm).

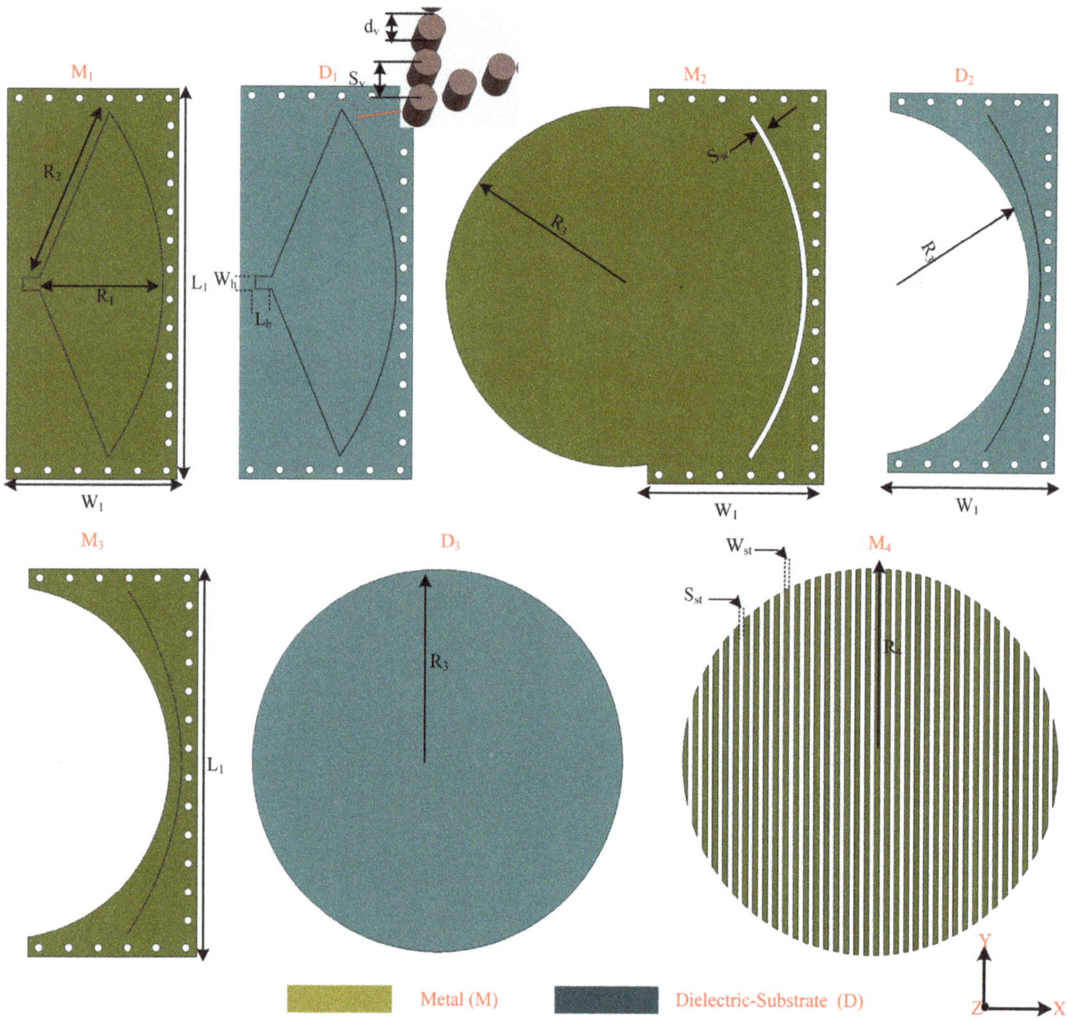

Figure 4. The layers' details of the proposed design, M and D, refer to the metal and dielectric layer, respectively. There are three dielectric layers and four metallic layers.

Table 1. Dimensions of the proposed antenna; all values in mm.

L_1	W_1	R_1	R_2	R_3	R_4	L_h	W_h	S_v	d_v	S_w	W_{st}	S_{st}
340	115	111.4	157.2	155	150	16	12	1.5	1	5	4	4

2.2.1. Radiating Disc

Linear periodic metallic patterns in a circular manner are printed on a disc to form a top layer, as depicted in Figure 4 (M_4). The disc is separated from the other structure to produce beam steering by its rotation (D_3 with M_4). The configuration of metallic patterns can be chosen based on the required beam direction; it works as an antenna array. Thus, the critical factors are the number of strips, distance between them, and width of each strip. The space between the elements is chosen to avoid any grating lobes and achieve high gain. By changing the shape and size of the metallic patterns, the beam can be varied. To demonstrate this idea, metallic patterns are printed on the top side of the disc. After careful optimization of the patterns, a period of 8 mm and a total length of 15 λ along the x-direction at 20 GHz are chosen. A width of 4 mm and a spacing of 4 mm, which provide beam pointing at the $phi = 0°$ and $theta = 8°$ direction, are considered.

2.2.2. Feeding Structure

To make the antenna compact, pillbox-based feeding is implemented underneath the disc, as shown in Figure 2. A detailed view of the feeding mechanism along with the used substrates is described in Figure 5. The feeding structure consists of an SIW horn feeding, parabolic reflector, and coupling slot, as shown in Figure 4. The geometric properties of the parabolic reflectors offer the transformation of a cylindrical wave originating from the paraboloid's focus into a plane wave directed along the parabola axis. Therefore, the feeding structure and the coupling slot are arranged and placed in different layers to avoid aperture blocking. The feeding is positioned in the focal plane of the parabolic reflector located on the lower substrate (D_1), which is integrated with the horn and SIW structure to feed and guide the waver, respectively. Meanwhile, the coupling slot is placed on the upper substrate (D_2). Connectivity between the two layers is facilitated through this coupling slot. The reflector surfaces are actualized using vertical metallic vias, establishing a connection between the top metal layers M_1 and M_3. Additionally, a coupling slot with a width of 5 mm is etched onto the middle metal layer (M_2). The SIW horn, which has dimensions of L_1 and L_2, is fed by the 2.92 mm millimeter-wave 50 Ω connector and generates a quasi-cylindrical wave (see Figure 6a). The wave propagates via substrate 3 (Figure 5) towards the parabolic reflector. The parabolic reflector is also SIW-based and converts the incoming wave into a plane wave. The parabolic reflector is generally designed using [34,41]:

$$r = \frac{2F}{1 + cos\phi} \tag{1}$$

where r is the parabolic surface radius, F refers to the focal length, and ϕ indicates the angle between r and F vectors (see Figure 5b). The parabolic slot is positioned close to the parabolic surface for maximum power coupling. The coupling slot transforms that plane wave to the top layer (substrate 1), which has metallic patterns; see Figure 6b. The coupling slot, which has a width of s_w, is placed just on the edge of the parabolic reflector. s_w is a critical parameter to improve the impedance-matching bandwidth. The optimized value of the s_w is 5 mm for the final design. The SIW horn and parabolic reflector are made of metallic vias that have a radius of 0.5 mm. The focal length and curve aperture are optimized to achieve a low side lobe level. The parabolic reflector has a curve aperture of 301 mm and focal point distance (F) of 126 mm.

Figure 6 shows the electric field distributions of the antenna for the top layer (radiating strips) and bottom layer (pillbox feeding) to help understand its working mechanism. Pillbox feeding generates a plane wave in the x-direction, which excites the radiating strips as a leaky-wave antenna. This excitation produces a broadside radiation pattern when there is no rotation in patterns. However, when the metallic patterns are rotated by an angle $\omega = \pm 20°$, the beam can be steered in azimuth and elevation planes. This can be observed from the electric field distributions in Figure 6c,d. Specifically, to achieve beam scanning, the intersection of the plane wave and the radiating metallic strips is changed by rotating the disc, as shown in Figure 6c,d. The disc is rotated (ω) clockwise and anti-clockwise to produce symmetrical beam scanning.

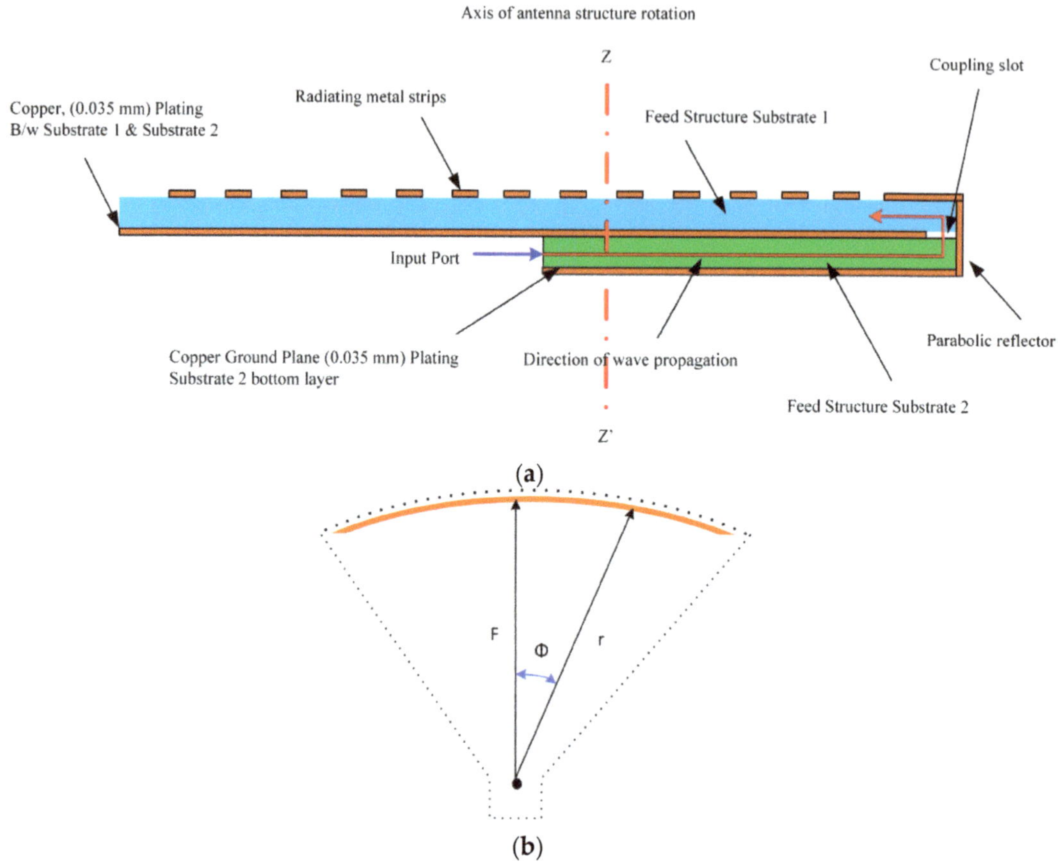

Figure 5. (**a**) Detailed view of layers and material stack up for the antenna and (**b**) pillbox schematic.

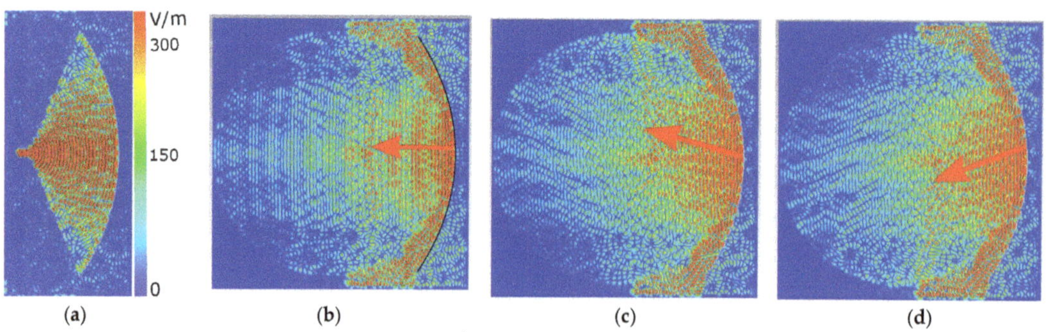

Figure 6. The electric field distributions obtained at 20 GHz. (**a**) The bottom layer and the top layer (**b**) without any rotation $\omega = 0°$, (**c**) at the maximum scanning angle on the right side $\omega = 20°$, and (**d**) at the maximum scanning angle at the left side $\omega = -20°$. Red arrows are pointing in the direction of the propagation of waves.

3. Antenna Performance and Discussions

A prototype was fabricated to validate the proposed idea. Figure 7 shows the photograph of the top and bottom views of the fabricated layers before the assembly of the proposed antenna. The three layers are fabricated as layer-1, which is the feeding parabolic, layer-2, which is the coupling layer, and layer-3, which is the radiating layer. It should be noted that the design consists of multi-layer substrates. Small holes are made in the corner of the substrates to assemble the whole structure via nylon screws. A 2.92 mm millimeter-wave 50 Ω connector is soldered from the bottom side of the antenna. Layer-1 and layer-2 were assembled together using nylon screws, and then the rotating disc was placed on the top side and rotated about its central axis to realize beam scanning, as shown in Figure 8a,b. After assembling the antenna, a vector network analyzer was used to test its reflection coefficients. The antenna's tested reflection coefficients are shown in Figure 9 in a blue-dotted–dashed line. The tested results are also compared with simulations, shown in Figure 9 as a red solid line. It can be seen from the measured results that the 10 dB return loss (−10 dB reflection coefficients) bandwidth is from 19.2 GHz to 20.4 GHz. The measured bandwidth is about 1.2 GHz, which is slightly smaller than the simulated one. The discrepancies between the measured and simulated reflection coefficients are mainly caused by fabrication, soldering, and measurement errors. It should be noted that the antenna has multiple layers and is assembled in the laboratory; this type of discrepancy is expected due to the manual assembly of the layers in addition to the human factor error of the soldering, where the soldering point can work as an additional stub at a higher frequency, which induces mismatch between the antenna and the connector. This can be minimized by having more accurate assembly of the layers, reducing the air gap between layers, and minimizing measurement errors.

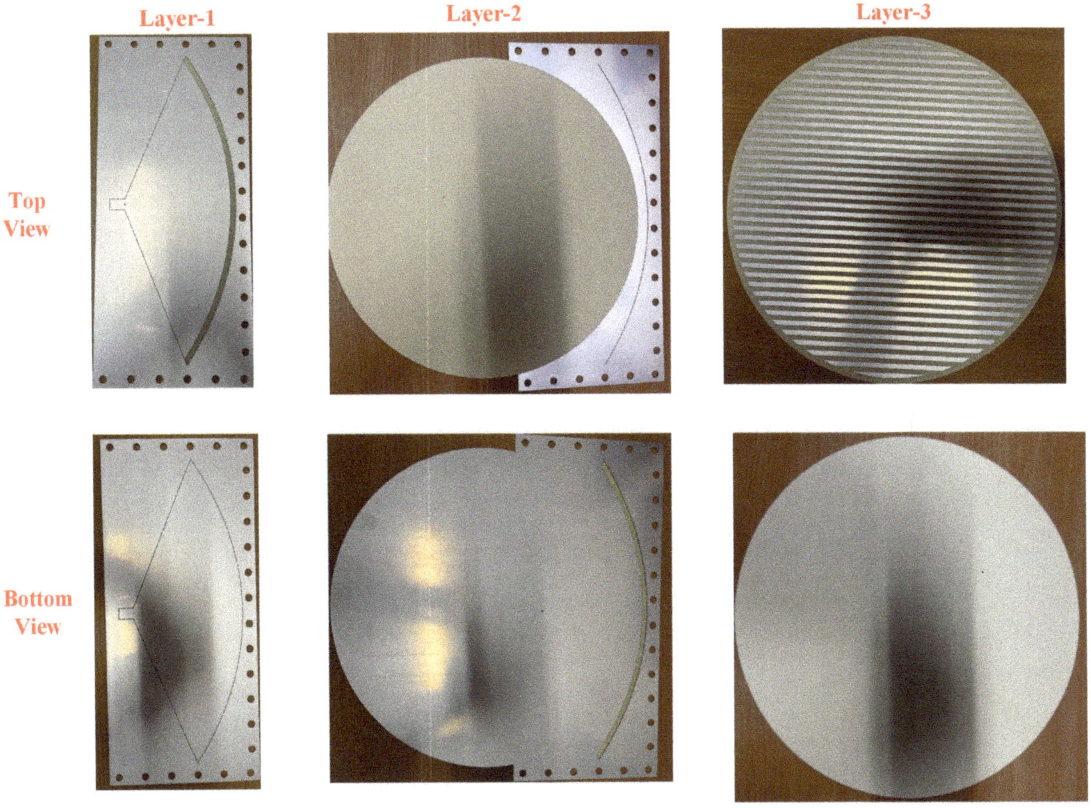

Figure 7. Photographs of the top and bottom views of the fabricated layers before the assembly. Layer-1 is the parabolic feed structure, layer-2 is the coupling layer, and layer-3 is the radiating strips layer.

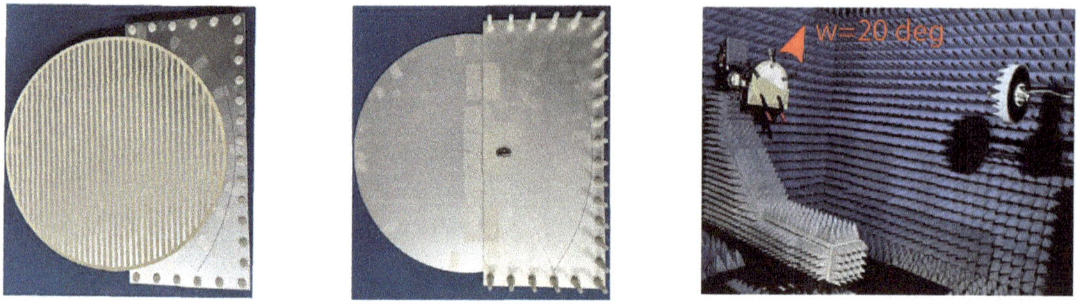

(**a**) (**b**) (**c**)

Figure 8. Photographs of the fabricated prototype. (**a**) Top view, (**b**) bottom views, and (**c**) measurement setup with a disc rotation of 20°.

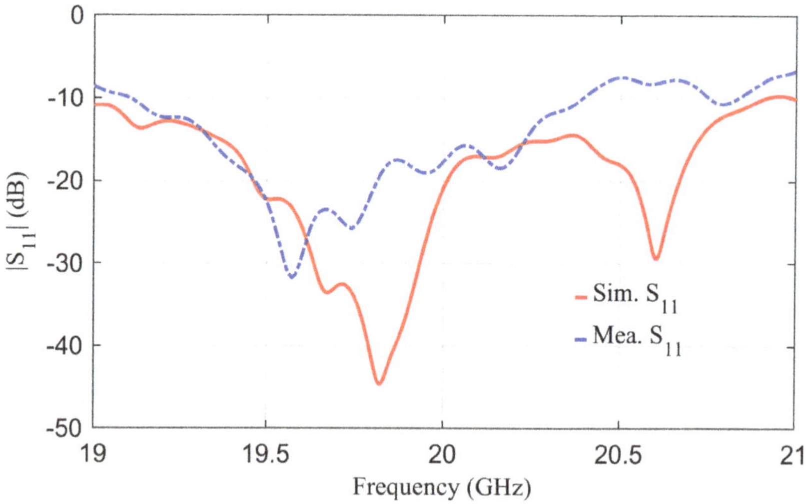

Figure 9. The simulated and measured reflection coefficient of the proposed antenna versus frequency.

The mechanical rotation is manually realized by rotation steps of $\omega = 5°$ from $-20°$ to $20°$. The measurement setup for $\omega = 20°$ is shown in Figure 8c. The antenna is tested for radiation patterns and realized gain values. To understand the beam steering of the proposed antenna, the normalized measured radiation patterns are shown in Figure 10. It is evident that the beam direction is steered with a little gain variation ($\sim$3 dBi) over the whole rotating angle. The radiation patterns of nine different rotation angles have been plotted in Figure 10 from $-20°$ to $20°$ with a step of $5°$. The intermediate coverage area between the beams can be covered by selecting small intermediate rotation angles. It is noted that the antenna offers $120°$ as a scanning angle, while Figure 11 depicts the 2D plots of the measured radiation patterns at 20 GHz for the nine rotation angles for the whole domain $0 \leq \theta \leq 180°; -180° \leq \phi \leq 180°$. The plots demonstrate the ability of the antenna to achieve beam steering with high gain and low side lobes. On the other hand, one of the essential merits of evaluating the scanning area of the antenna is the total scanning angle calculated from the total scan pattern (TSP) [42–44]. The TSP is the maximum achievable gain at each certain angular angle (θ, ϕ) in the whole spherical coordinate, where the TSP is computed for all potential antenna rotation angles $(\omega_1 : \omega_n)$ and then the maximum gain (G) is selected at each point in the spatial distribution as [44]:

$$TSP(\theta,\ \phi) = \max\left[G_{\omega_1}(\theta,\ \phi),\ G_{\omega_2}(\theta,\ \phi),\ \ldots\ldots\ldots,\ G_{\omega_n}(\theta,\ \phi)\right],\ 0 \le \theta \le 180^\circ; -180^\circ \le \phi \le 180^\circ \quad (2)$$

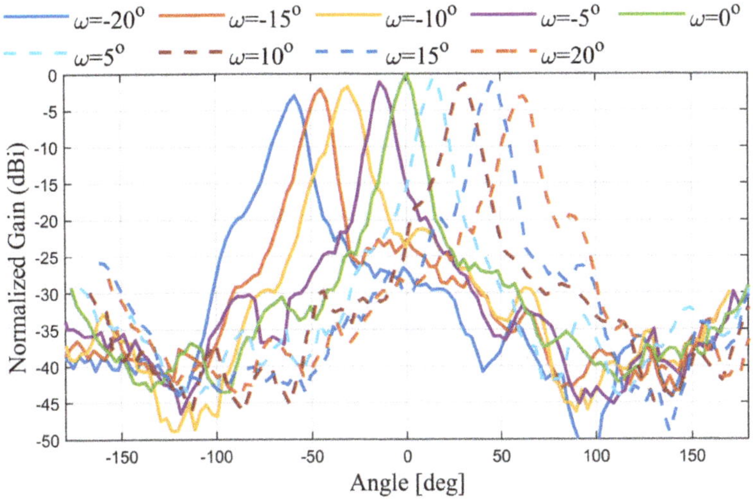

Figure 10. The measured radiation patterns of the antenna with the rotation angles from -20° to 20°. All the radiation patterns are normalized to the maximum gain at $\omega = 0^\circ$.

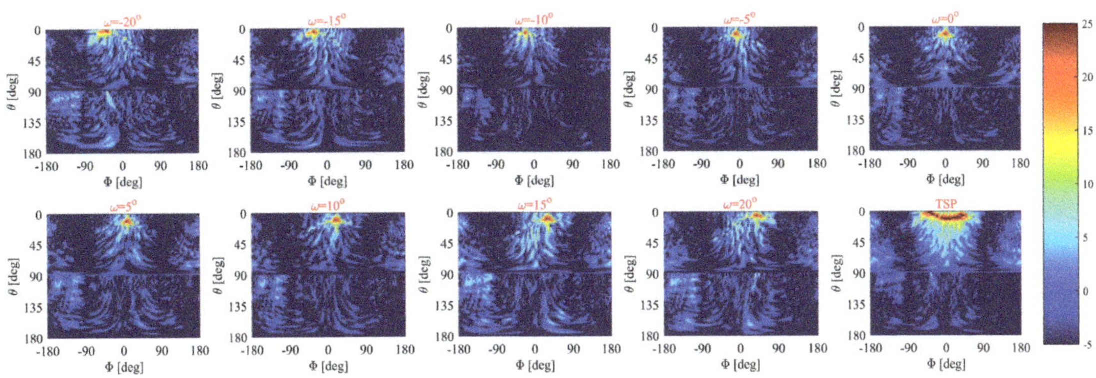

Figure 11. Two-dimensional plot of measured radiation patterns in dBi at rotation angles from -20° to 20° (with 5° spatial angle) and the total scan pattern of those rotation angles. All the plotted values present the gain at a whole spherical angle for each rotated angle, and then TSP is calculated based on (2).

As shown in Figure 11, the measured $TSP(\theta,\ \phi)$ demonstrates the ability of the antenna to achieve beam steering of 120° with a tiny variation in the beam in the elevation plane ($\pm 4^\circ$) and without any null, which assembles only from nine rotations of the disc within $\pm 20^\circ$. The simulated and measured gain values of the steering beam at the different rotating angles are shown in Table 2. The maximum measured gain is 23.2 dBi at $w = 0^\circ$ rotation, which is 3.3 dB less than the simulated one. This difference in gain values is expected because of higher losses of the substrate at a high frequency. It is also attributed to the fabrication, soldering, air gaps, and measurement errors. Nevertheless, this is acceptable for large antennas, especially those operating at high-frequency bands. The variations in the measured gain from -40° to $+40^\circ$ scanning range is only 2 dBi, which is better than the other reported designs [21,36], while a 3.3 dBi variation is observed in the whole scanning range from −60° to +60°.

Table 2. The simulated and measured peak gain at 20 GHz for different rotation angles.

Angle of Rotation ω	**−20°**	**−10°**	**0°**	**10°**	**20°**
Simulated Gain (dBi)	24	25.5	26.5	25.5	24
Measured Gain (dBi)	20.21	22.1	23.2	22.5	20.18

In order to determine the quality of the antenna and its performance, it is compared to other reference antennas in terms of antenna height, frequency band, beam coverage, measured gain, and SLLs at the broadside and maximum scan angle. These characteristics are given in Table 3. By rotating a horn within the focal plane of the pillbox feeding system, the reference antenna that was described in [39] possesses a high gain and low SLL when viewed from the broadside. However, it displays a high SLL when viewed from a maximum scan angle of ±40 degrees. Additionally, the antenna height is $4.39\lambda_0$ which is much higher than the proposed design. Furthermore, it features a mechanically steered main beam that extends beyond ±40 degrees. The other reference antenna that is presented in [21] has a low SLL at the broadside and a maximum scan angle of 40 degrees; however, the measured gain is low (i.e., 19.6 dBi), and the beam coverage that is achieved is only ±40 degrees. Although the antennas in [35,36,40] offer a small profile, they suffer from either a limited coverage angle or high side lobe level. The antennas in [39,45] have a wider bandwidth compared to the others due to using high profiles (4.39 λ_0 and 9 λ_0). However, they offer limited angle scanning and high SLL. The antenna proposed in this study, on the other hand, possesses a relatively high gain, which is, i.e., 23.2 dBi. Additionally, it has a low SLL of −17.8 dB at a maximum scan angle of 60 degrees. Furthermore, the antenna height is only 0.3 λ_0, and it has a beam coverage of ±60 degrees on the azimuth plane.

Table 3. Performance comparison of the proposed and previous quasi-optical antenna system.

Ref.	**Antenna Height λ_0**	**BW (GHz)**	**Beam Coverage (Degree)**	**Measured Gain (dBi)**	**SLL (dB)**	
					@ BS	**@ MSA**
[35]	0.08	23.9–24.9	±36	17.7	−15	−8.8
[40]	0.08	23.5–25.7	±30	23.8	−12	−16
[39]	4.39	27.5–31.0	±40	28.9	−32	−11
[36]	0.08	23.5–24.6	±40	25	−15	−10
[21]	0.32	10.2–13.9	±40	19.6	−30	−15.6
[45]	9	28.5–31.0	±24	31.3	−20	−10
[46]	0.1	23.8–24.3	±39	21.6	−23	−10
[47]	0.1	23.8–24.4	±40	24.2	−24	−11
[48]	0.1	23.5–24.5	±35	22	−20	−15
[49]	0.25	74–78	±40	24	−25	−10
This Work	0.3	19.0–20.4	±60	23.2	−20	−17.8

SLL: side lobe level, BW: bandwidth, BS broadside angle, MSA: maximum scanning angle.

4. Conclusions

In this work, wide-angle beam steering has been shown for the purpose of on-the-move satellite communication through the utilization of an innovative closed-form pillbox antenna system. The complete wide-angle beam steering antenna system includes three main parts: fixed-feed SIW-based horn, novel quasi-optical system, and leaky-wave radiating part. A transition made by a single coupling slot alongside and without spacing from the parabolic reflector has been proposed for the quasi-optical system to have effective coupling and extend the bandwidth. Through the utilization of the mechanical rotation of the metallic patterns, beam scanning is accomplished. The experimental results show that the antenna can provide a beam-scanning capability of 120° in the azimuth plane. The antenna terminal demonstrates directive radiation pasterns with a maximum measured gain of 23.2 dBi with an antenna height of $0.3\lambda_0$ and low SLL of -17.8 dB at a maximum scan angle of 60°. The antenna operates in the frequency range from 19.2 GHz to 20.4 GHz with a reflection coefficient less than -10 dB. This concept has the potential to be extended to achieve circular polarization, which might be accomplished by incorporating a superstrate layer of a thin polarizer.

Author Contributions: Conceptualization, M.I., A.T.M. and A.A.; design, M.I. and K.S.; software, K.S.; validation, M.I. and M.M.; analysis, K.S. and M.I; investigation, A.T.M. and M.M.; writing—original draft preparation, M.I. and K.S.; writing—review and editing, A.T.M., M.M. and A.A.; supervision A.A. All authors have read and agreed to the published version of the manuscript.

Funding: This research was supported by the Australian Research Council under ARC Grant LP160100917.

Data Availability Statement: All data has been included in the paper.

Conflicts of Interest: The authors declare no conflicts of interest.

References

1. Ionescu, L.; Rusu-Casandra, A.; Bira, C.; Tatomirescu, A.; Tramandan, I.; Scagnoli, R.; Istriteanu, D.; Popa, A.-E. Development of the Romanian Radar Sensor for Space Surveillance and Tracking Activities. *Sensors* **2022**, *22*, 3546. [CrossRef] [PubMed]
2. Gui, G.; Liu, M.; Tang, F.; Kato, N.; Adachi, F. 6G: Opening New Horizons for Integration of Comfort, Security, and Intelligence. *IEEE Wirel. Commun.* **2020**, *27*, 126–132. [CrossRef]
3. Zhou, H.; Pal, A.; Mehta, A.; Nakano, H.; Modigliana, A.; Arampatzis, T.; Howland, P. Reconfigurable Phased Array Antenna Consisting of High-Gain High-Tilt Circularly Polarized Four-Arm Curl Elements for Near Horizon Scanning Satellite Applications. *IEEE Antennas Wirel. Propag. Lett.* **2018**, *17*, 2324–2328. [CrossRef]
4. Bariah, L.; Mohjazi, L.; Muhaidat, S.; Sofotasios, P.C.; Kurt, G.K.; Yanikomeroglu, H.; Dobre, O.A. A Prospective Look: Key Enabling Technologies, Applications and Open Research Topics in 6G Networks. *IEEE Access* **2020**, *8*, 174792–174820. [CrossRef]
5. Chu, D.; Mao, Y.; Li, H.; Bie, H.; Zhou, Y. Dual-Polarized Multi-Beam Fixed-Frequency Beam Scanning Leaky-Wave Antenna. *Sensors* **2023**, *23*, 5070. [CrossRef] [PubMed]
6. Aljaloud, K.; Sultan, K.; Ikram, M.; Alqahtani, A.H.; Abbasi, Q.H.; Hussain, R. Low-Profile Antenna System for Cognitive Radio in IoST CubeSat Applications. *Sensors* **2023**, *23*, 4782. [CrossRef]
7. Soumya, A.; Mohan, C.K.; Cenkeramaddi, L.R. Recent Advances in mmWave-Radar-Based Sensing, Its Applications, and Machine Learning Techniques: A Review. *Sensors* **2023**, *23*, 8901. [CrossRef]
8. Moon, S.-M.; Cho, J.; Lee, H.L. High Gain Flat-Panel mmWave Antenna Array. *Sensors* **2023**, *23*, 9433. [CrossRef]
9. Merino-Fernandez, I.; Khemchandani, S.L.; del Pino, J.; Saiz-Perez, J. Phased Array Antenna Analysis Workflow Applied to Gateways for LEO Satellite Communications. *Sensors* **2022**, *22*, 9406. [CrossRef]
10. Elrahman, S.I.A.; Elkhawaga, A.M.; Hussein, A.H.; Shaalan, A.E.A. Linear Antenna Array Sectorized Beam Scanning Approaches Using Element Position Perturbation in the Azimuth Plane. *Sensors* **2023**, *23*, 6557. [CrossRef]
11. Ikram, M.; Sultan, K.S.; Abbosh, A.M.; Nguyen-Trong, N. Sub-6 GHz and mm-Wave 5G Vehicle-to-Everything (5G-V2X) MIMO Antenna Array. *IEEE Access* **2022**, *10*, 49688–49695. [CrossRef]
12. Zetterstrom, O.; Hamarneh, R.; Quevedo-Teruel, O. Experimental Validation of a Metasurface Luneburg Lens Antenna Implemented With Glide-Symmetric Substrate-Integrated Holes. *IEEE Antennas Wirel. Propag. Lett.* **2021**, *20*, 698–702. [CrossRef]
13. Sultan, K.; Abdullah, H.; Abdallah, E.; El-Hennawy, H. MOM/GA-Based Virtual Array for Radar Systems. *Sensors* **2020**, *20*, 713. [CrossRef] [PubMed]
14. Sun, G.-H.; Wong, H. Millimeter-Wave High-Gain Magneto-Electric Dipole Antenna Array with Pillbox Corporate Feed Network. *IEEE Trans. Antennas Propag.* **2021**, *69*, 5631–5639. [CrossRef]
15. Xu, R.; Gao, S.S.; Li, J.; Wei, K.; Luo, Q. A Reconfigurable Dual-Band Dual-Circularly Polarized Antenna for Vehicle Global Navigation Satellite System Application. *IEEE Trans. Veh. Technol.* **2020**, *69*, 11857–11867. [CrossRef]
16. Yang, H.-Z.; Qu, S.-W. High-Gain Millimeter-Wave Beam Scanning Transmitarray Antenna. *Sensors* **2023**, *23*, 4709. [CrossRef]
17. Wen, L.; Gao, S.; Luo, Q.; Hu, W.; Yin, Y. Wideband Dual Circularly Polarized Antenna for Intelligent Transport Systems. *IEEE Trans. Veh. Technol.* **2020**, *69*, 5193–5202. [CrossRef]
18. Xu, J.; Hong, W.; Jiang, Z.H.; Zhang, H. Low-Profile Circular Patch Array Fed by Slotted Substrate Integrated Waveguide. *IEEE Trans. Antennas Propag.* **2019**, *67*, 960–970. [CrossRef]
19. Hamberger, G.F.; Spath, S.; Siart, U.; Eibert, T.F. A Mixed Circular/Linear Dual-Polarized Phased Array Concept for Automotive Radar—Planar Antenna Designs and System Evaluation at 78 GHz. *IEEE Trans. Antennas Propag.* **2019**, *67*, 1562–1572. [CrossRef]
20. He, G.; Gao, X.; Sun, L.; Zhang, R. A Review of Multibeam Phased Array Antennas as LEO Satellite Constellation Ground Station. *IEEE Access* **2021**, *9*, 147142–147154. [CrossRef]

21. Lou, T.; Yang, X.-X.; Cao, Q.-D.; Gao, S. A Low Profile Circularly Polarized Beam Scanning Patch Array Fed by Parallel Plate Waveguide. *IEEE Trans. Antennas Propag.* **2022**, *70*, 7384–7392. [CrossRef]
22. Chen, C.-N.; Hung, L.-C.; Tang, T.-C.; Chao, W.-P.; Chen, C.-Y.; Chuang, P.-H.; Lin, G.-Y.; Liao, W.-J.; Nien, Y.-H.; Huang, W.-C.; et al. 38-GHz Phased Array Transmitter and Receiver Based on Scalable Phased Array Modules with Endfire Antenna Arrays for 5G MMW Data Links. *IEEE Trans. Microw. Theory Tech.* **2021**, *69*, 980–999. [CrossRef]
23. Gray, D.; Lu, J.W.; Thiel, D. Electronically steerable Yagi-Uda microstrip patch antenna array. *IEEE Trans. Antennas Propag.* **1998**, *46*, 605–608. [CrossRef]
24. Maqsood, M.; Gao, S.; Brown, T.W.C.; Unwin, M.; Steenwijk, R.d.V.V.; Xu, J.D.; Underwood, C.I. Low-Cost Dual-Band Circularly Polarized Switched-Beam Array for Global Navigation Satellite System. *IEEE Trans. Antennas Propag.* **2014**, *62*, 1975–1982. [CrossRef]
25. Chaloun, T.; Boccia, L.; Arnieri, E.; Fischer, M.; Valenta, V.; Fonseca, N.J.G.; Waldschmidt, C. Electronically Steerable Antennas for Future Heterogeneous Communication Networks: Review and Perspectives. *IEEE J. Microwaves* **2022**, *2*, 545–581. [CrossRef]
26. Zhang, Y.; Han, Z.; Tang, S.; Shen, S.; Chiu, C.-Y.; Murch, R. A Highly Pattern-Reconfigurable Planar Antenna With 360° Single- and Multi-Beam Steering. *IEEE Trans. Antennas Propag.* **2022**, *70*, 6490–6504. [CrossRef]
27. Hu, W.; Arrebola, M.; Cahill, R.; Encinar, J.A.; Fusco, V.; Gamble, H.S.; Alvarez, Y.; Las-Heras, F. 94 GHz Dual-Reflector Antenna with Reflectarray Subreflector. *IEEE Trans. Antennas Propag.* **2009**, *57*, 3043–3050. [CrossRef]
28. Eid, A.; Hester, J.G.D.; Tentzeris, M.M. Rotman Lens-Based Wide Angular Coverage and High-Gain Semipassive Architecture for Ultralong Range mm-Wave RFIDs. *IEEE Antennas Wirel. Propag. Lett.* **2020**, *19*, 1943–1947. [CrossRef]
29. Gao, Y.; Khaliel, M.; Zheng, F.; Kaiser, T. Rotman Lens Based Hybrid Analog–Digital Beamforming in Massive MIMO Systems: Array Architectures, Beam Selection Algorithms and Experiments. *IEEE Trans. Veh. Technol.* **2017**, *66*, 9134–9148. [CrossRef]
30. Mirmozafari, M.; Tursunniyaz, M.; Luyen, H.; Booske, J.H.; Behdad, N. A Multibeam Tapered Cylindrical Luneburg Lens. *IEEE Trans. Antennas Propag.* **2021**, *69*, 5060–5065. [CrossRef]
31. Ao, T.; Pan, Y.; Dong, Y. Low-Profile Dual-Polarized Luneburg Lens Based on TE/TM Surface Wave Modes. *IEEE Antennas Wirel. Propag. Lett.* **2022**, *21*, 1862–1866. [CrossRef]
32. Paul, T.; Harinath, M.; Garg, S.K.; Aich, S.; Kumar, A.; Trivedi, J.; Kumar, A.; Patel, M.K.; Rao, C.V.N.; Jyoti, R. Miniaturized High-Power Beam Steering Network Using Novel Nonplanar Waveguide Butler Matrix. *IEEE Microw. Wirel. Components Lett.* **2021**, *31*, 678–681. [CrossRef]
33. Shao, Q.; Chen, F.-C. Design of 2 × 8 Filtering Butler Matrix with Arbitrary Power Distribution. *IEEE Trans. Circuits Syst. II Express Briefs* **2021**, *68*, 3527–3531. [CrossRef]
34. Sun, G.-H.; Wong, H. A Planar Millimeter-Wave Antenna Array with a Pillbox-Distributed Network. *IEEE Trans. Antennas Propag.* **2020**, *68*, 3664–3672. [CrossRef]
35. Cao, Y.; Yan, S.; Li, J.; Chen, J. A Pillbox Based Dual Circularly-Polarized Millimeter-Wave Multi-Beam Antenna for Future Vehicular Radar Applications. *IEEE Trans. Veh. Technol.* **2022**, *71*, 7095–7103. [CrossRef]
36. Tekkouk, K.; Ettorre, M.; Le Coq, L.; Sauleau, R. SIW Pillbox Antenna for Monopulse Radar Applications. *IEEE Trans. Antennas Propag.* **2015**, *63*, 3918–3927. [CrossRef]
37. Wang, C.; Yao, Y.; Cheng, X.; Zhu, Z.; Li, X. A W-Band High-Efficiency Multibeam Circularly Polarized Antenna Array Fed by GGW Butler Matrix. *IEEE Antennas Wirel. Propag. Lett.* **2021**, *20*, 1130–1134. [CrossRef]
38. Lee, J.-G.; Lee, J.-H. Low-Profile High-Efficiency Transmitarray Antenna for Beamforming Applications. *Electronics* **2023**, *12*, 3178. [CrossRef]
39. Ettorre, M.; Manzillo, F.F.; Casaletti, M.; Sauleau, R.; Le Coq, L.; Capet, N. Continuous Transverse Stub Array for Ka-Band Applications. *IEEE Trans. Antennas Propag.* **2015**, *63*, 4792–4800. [CrossRef]
40. Ettorre, M.; Sauleau, R.; Le Coq, L. Multi-Beam Multi-Layer Leaky-Wave SIW Pillbox Antenna for Millimeter-Wave Applications. *IEEE Trans. Antennas Propag.* **2011**, *59*, 1093–1100. [CrossRef]
41. Balanis, C.A. *Antenna Theory: Analysis and Design*, 4th ed.; Wiley: Hoboken, NJ, USA, 2016.
42. Sultan, K.; Ikram, M.; Nguyen-Trong, N. A Multiband Multibeam Antenna for Sub-6 GHz and mm-Wave 5G Applications. *IEEE Antennas Wirel. Propag. Lett.* **2022**, *21*, 1278–1282. [CrossRef]
43. Di Paola, C.; Zhao, K.; Zhang, S.; Pedersen, G.F. SIW Multibeam Antenna Array at 30 GHz for 5G Mobile Devices. *IEEE Access* **2019**, *7*, 73157–73164. [CrossRef]
44. Di Paola, C.; Zhang, S.; Zhao, K.; Ying, Z.; Bolin, T.; Pedersen, G.F. Wideband Beam-Switchable 28 GHz Quasi-Yagi Array for Mobile Devices. *IEEE Trans. Antennas Propag.* **2019**, *67*, 6870–6882. [CrossRef]
45. Smierzchalski, M.; Manzillo, F.F.; Del Mastro, M.; Capet, N.; Palacin, B.; Sauleau, R.; Ettorre, M. A Novel Dual-Polarized Continuous Transverse Stub Antenna Based on Corrugated Waveguides—Part II: Experimental Demonstration. *IEEE Trans. Antennas Propag.* **2021**, *69*, 1313–1323. [CrossRef]
46. Tekkouk, K.; Ettorre, M.; Sauleau, R. Multibeam Pillbox Antenna Integrating Amplitude-Comparison Monopulse Technique in the 24 GHz Band for Tracking Applications. *IEEE Trans. Antennas Propag.* **2018**, *66*, 2616–2621. [CrossRef]
47. Tekkouk, K.; Ettorre, M.; Gandini, E.; Sauleau, R. Multibeam Pillbox Antenna with Low Sidelobe Level and High-Beam Crossover in SIW Technology Using the Split Aperture Decoupling Method. *IEEE Trans. Antennas Propag.* **2015**, *63*, 5209–5215. [CrossRef]

8. Gandini, E.; Ettorre, M.; Casaletti, M.; Tekkouk, K.; Le Coq, L.; Sauleau, R. SIW Slotted Waveguide Array with Pillbox Transition for Mechanical Beam Scanning. *IEEE Antennas Wirel. Propag. Lett.* **2012**, *11*, 1572–1575. [CrossRef]
9. Ettorre, M.; Sauleau, R.; Le Coq, L.; Bodereau, F. Single-Folded Leaky-Wave Antennas for Automotive Radars at 77 GHz. *IEEE Antennas Wirel. Propag. Lett.* **2010**, *9*, 859–862. [CrossRef]

sensors

MDPI

Article

A Design Method and Application of Meta-Surface-Based Arbitrary Passband Filter for Terahertz Communication

Da Hou [1,2], Lihui Wang [1], Qiuhua Lin [1], Xiaodong Xu [2,3], Yin Li [2], Zhiyong Luo [1,2,4,*] and Hao Chen [2,*]

1 School of Electronics and Communication Engineering, Sun Yat-sen University, Shenzhen 518107, China; houd5@mail2.sysu.edu.cn (D.H.); wanglh65@mail2.sysu.edu.cn (L.W.); linqh26@mail2.sysu.edu.cn (Q.L.)
2 Pengcheng Laboratory, Shenzhen 518107, China; xuxiaodong@bupt.edu.cn (X.X.); liy17@pcl.ac.cn (Y.L.)
3 School of Information and Communication Engineering, Beijing University of Posts and Telecommunications, Beijing 100876, China
4 Shenzhen Key Laboratory of Navigation and Communication Integration, Shenzhen 518107, China
* Correspondence: luozhy57@mail.sysu.edu.cn (Z.L.); chenh03@pcl.ac.cn (H.C.)

Abstract: A meta-surface-based arbitrary bandwidth filter realization method for terahertz (THz) future communications is presented. The approach involves integrating a meta-surface-based band-stop filter into an ultra-wideband (UWB) bandpass filter and adjusting the operating frequency range of the meta-surface bandstop filter to realize the design of arbitrary bandwidth filters. It effectively addresses the complexity of designing traditional arbitrary bandwidth filters and the challenges in achieving impedance matching. To underscore its practicality, the paper employs silicon substrate integrated gap waveguide (SSIGW) and this method to craft a THz filter. To begin, design equations for electromagnetic band gap (EBG) structures were developed in accordance with the requirements of through-silicon via (TSV) and applied to the design of the SSIGW. Subsequently, this article employs equivalent transmission line models and equivalent circuits to conduct theoretical analyses for both the UWB passband and the meta-surface stopband portions. The proposed THz filter boasts a center frequency of 0.151 THz, a relative bandwidth of 6.9%, insertion loss below 0.68 dB, and stopbands exceeding 20 GHz in both upper and lower ranges. The in-band group delay is 0.119 ± 0.048 ns. Compared to reported THz filters, the SSIGW filter boasts advantages such as low loss and minimal delay, making it even more suitable for future wireless communication.

Keywords: terahertz; silicon substrate integrated gap waveguide (SSIGW); through-silicon via (TSV); meta-surface; filter

1. Introduction

The "International Mobile Telecommunications (IMT) Framework and General Objectives Proposal for 2030 and Future Development" program document's completion in 2023 ushers in a new phase of future wireless communications development. Terahertz (THz) technology, one of the primary future wireless communication technologies, will be extensively used in both the civil and military sectors. As an indispensable component in communication systems, filters play a crucial role in the efficient and low-loss design of the THz band. However, THz filters, recognized as pivotal components, currently grapple with challenges such as significant losses, integration complexities, and difficulties in achieving arbitrary bandwidth designs. The current implementation of THz filters relies mainly on meta-material design [1,2], making it challenging to integrate them into silicon-based front-end circuits. Some waveguide-based filters' [3,4] design faces issues such as surface waves and high interconnection losses.

In 2015, substrate integrated gap waveguide (SIGW) was proposed to solve the problems of loss and packaging in the design of high-frequency devices [5]. SIGW has the capability to mitigate surface waves and space radiation in transmission lines, providing a favorable environment for millimeter-wave (MMW) circuits and their packaging [6–8].

However, when SIGW is applied in the THz range, the waveguide structures become extremely small, making them challenging to manufacture. Additionally, issues such as high losses and difficulties in integration arise. Fortunately, through-silicon via (TSV), a key semiconductor technology of 3D integrated circuits, has high-precision trench etching and metal filling on the order of microns and very low losses, which has aroused significant research interest [9]. Moreover, TSV can be connected to other planar devices through a redistribution layer [10]. Therefore, those attractive features have made TSV technology a good alternative for miniaturization and integration of passive devices. Indeed, some passive devices based on TSV have been proposed [3,11]. Recently, due to the large-scale expansion of semiconductor technology and emerging markets such as automotive radar and 5G, TSV technology has been gradually extended to MMW and THz fields [12–14]. In 2022, a silicon substrate integrated gap waveguide (SSIGW) utilizing TSV was proposed to address the aforementioned issues, and it has been successfully applied in the design of a 6G filtering metalens antenna [15]. Therefore, SSIGW, characterized by low losses, compact dimensions, suppression of surface waves, and ease of integration, emerges as a superior alternative for THz circuit design.

Additionally, regarding filter design, it is common for designs to target specific frequency bands and ranges, with fixed structures, making it challenging to achieve arbitrary bandwidths. Currently, only a few studies have addressed this issue [16,17], all adopting the approach of increasing resonance points, as illustrated in Figure 1, which depicts the basic principle. One study proposed a filter prototype synthesis method for designing filters with arbitrary bandwidths [16]. Another study presented a method based on the theory of small reflections to design microstrip filters with arbitrary passbands [17]. It introduced a novel stub-loaded microstrip filter and derived an approximate reflection coefficient function for the input of the filter. However, these methods face difficulties in computation and complexities in design. Due to its simplicity in design and controllable parameters, electromagnetic meta-surfaces have been widely used in device design [18,19]. Based on the meta-surface resonance theory, this paper proposes a method of utilizing meta-surfaces in the design of ultra-wideband (UWB) filters, offering a solution for achieving filters with arbitrary bandwidths. And the proposed method offers advantages such as simplicity in implementation and high flexibility.

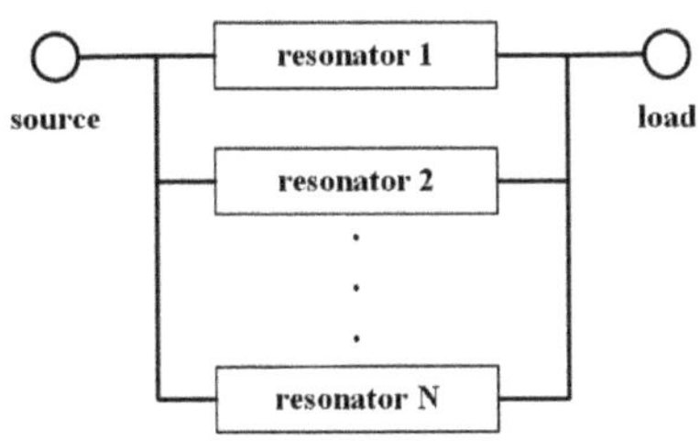

Figure 1. Diagram of the traditional implementation principle of arbitrary bandwidth filters.

This work proposes a method for the design of meta-surface filters with arbitrary passband, and for verification, a filter based on SSIGW is presented. In Section 2, the method for the design of meta-surface filters with arbitrary passband is given. Furthermore, an SSIGW meta-surface filter based on the design method has been proposed. In Section 3, a comprehensive analysis of the proposed filter has been undertaken, covering the theoretical design of SSIGW, the design of the UWB filter, and the meta-surface-based wideband bandstop filter design. Comparative evaluations with relevant THz filters have highlighted its superior performance. In Section 4, some conclusions are given.

2. Meta-Surface-Based Arbitrary Bandwidth SSIGW Filter Design Method

This section presents a method for designing an arbitrary bandwidth SSIGW filter based on meta-surfaces. The principle of the proposed method is illustrated in Figure 2. It entails incorporating a meta-surface-based bandstop filter into a UWB bandpass filter, with the controllable filter bandwidth achieved through the adjustment of the meta-surface's dimensions. The decision to employ meta-surface design for the bandstop structure is motivated by the challenges and complexity associated with achieving designs for diverse frequency bands using traditional bandstop filter structures. Furthermore, achieving perfect impedance matching with UWB filter structures has proven difficult using conventional band-stop filter configurations. Meta-surfaces offer advantages such as structural simplicity and flexibility, enabling the flexible design of filters with varying bandwidths and stopband widths to meet specific requirements. Figure 3 illustrates the theoretical performance of the proposed method. In Figure 3a, the performance of the bandpass filter is shown, and in Figure 3b, the performance of the bandstop filter is depicted. By superimposing these two, the desired filter can be obtained, as shown in Figure 3c.

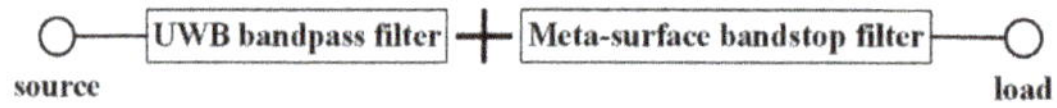

Figure 2. The proposed diagram for the implementation principle of arbitrary bandwidth filters.

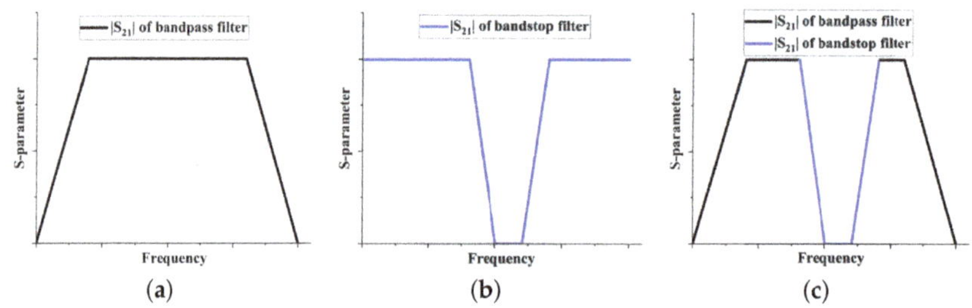

Figure 3. Theoretical performance of arbitrary bandwidth filter: (**a**) bandpass filter; (**b**) bandstop filter; and (**c**) required filter.

To validate the efficacy of the proposed approach, an SSIGW THz filter was designed employing split-ring resonator (SRR) meta-surfaces, as depicted in the overall structure illustrated in the Figure 4. It comprises two components: a UWB bandpass filter and a meta-surface wideband bandstop filter. Additionally, the meta-surface wideband bandstop filter is composed of meta-surface bandstop filter 1 and meta-surface bandstop filter 2. To provide a clearer exposition of the design concept, the subsequent sections will individually analyze the design of SSIGW, UWB filters, and meta-surface-based wideband bandstop filters.

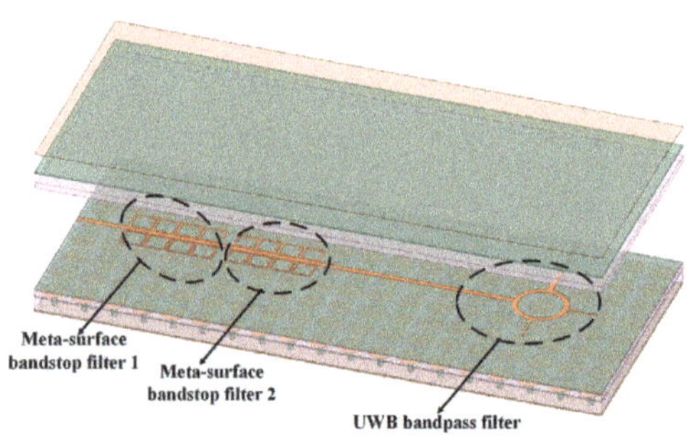

Figure 4. The physical structure of the proposed filter.

3. Physical Realization of Arbitrary Bandwidth SSIGW Filter Based on Meta-Surface

According to the literature [20], the sixth-generation mobile communication technology (6G) band includes the G band (0.14–0.22 THz). Therefore, this paper presents the detailed design of the filter in this frequency band, aiming to offer new insights into the filter design for 6G communication and promote the development of 6G communication. The concrete implementation of the proposed arbitrary bandwidth SSIGW filter based on meta-surface encompasses three key components: the theoretical design of SSIGW, the design of the ultra-wideband filter, and the design of the meta-surface-based band-stop filter with a wide stopband. The following subsections will delve into each of these aspects individually.

3.1. Design of SSIGW

Beginning with TSV technology, this subsection researches the design of SSIGW for the THz band. The dielectric constants of the materials employed in all structures within this artical are Si with $\varepsilon_s = 11.9$, and SiO_2 with $\varepsilon_o = 4$.

As per the information presented in [15], SSIGW is composed of a TSV-based electromagnetic band gap (EBG) array structure and a stripline. In this subsection, the TSV-based EBG structure is first given and analyzed so that an SSIGW can work for the THz band. TSV is formed in a silicon substrate with limited and nonzero resistivity and has a metal–insulator–semiconductor (MIS) structure [21], as shown in Figure 5, i.e., metal (copper), insulator (SiO_2), and semiconductor substrate (Si) between signal and ground [22]. The thickness of the insulator layer (T) is determined by the following formulae:

$$T > r\left(1 - e^{-\frac{2\pi\varepsilon_0 H}{\mathrm{Cox}(f)}}\right) \tag{1}$$

and

$$T > 0.2\ \mu\mathrm{m} \tag{2}$$

in which T is the thickness of the insulator SiO_2 layer outside the copper/silicon pillar; r is the radius of the copper column; H is the height of the copper column; and $Cox(f)$ is the maximum capacitance of the sidewall, which can be obtained by simulation.

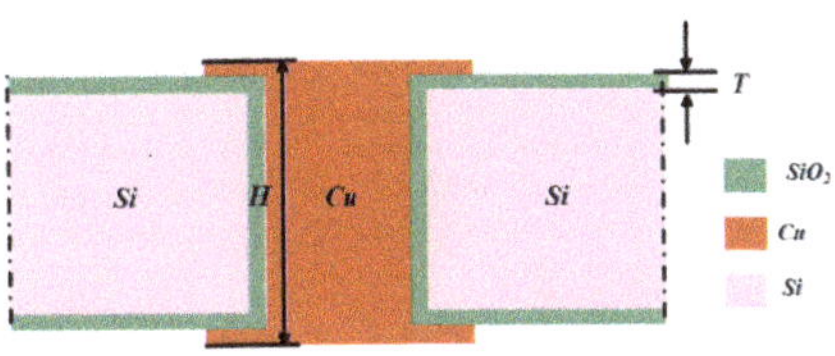

Figure 5. Schematic diagrams of a TSV cell.

Figure 6a shows the side view of TSV-based EBGs with indicated parameters, such as dimensions d_2, R, p, r, T, and g. Due to the computational difficulty of embedding the TSV-based EBG in two media, for simplicity, we assume that the TSV-based EBG is embedded in a medium with an equivalent relative permittivity, as shown in Figure 6b. Figure 6b shows the side view of TSV-based EBGs' equivalent structure with indicated parameters, such as capacitance C_1 and C_2, inductance L, and dimensions $d_1 = h_1 + 2T$, $d_2 = h_2 + h_3 + 2T$. The effective relative dielectric constants ε_{eff1} and ε_{eff2} are obtained from (3) and (4) using the weighted average method.

$$\varepsilon_{eff1} = \frac{2T\varepsilon_o + (p^2 - \pi T^2)h_1\varepsilon_s/p^2}{d_1} + \frac{\pi T^2 h_1 \varepsilon_o/p^2}{d_1} \tag{3}$$

$$\varepsilon_{eff2} = \frac{2T\varepsilon_o + (h_2 + h_3)\varepsilon_s}{d_2} \tag{4}$$

Applying the theory in [23,24] to Figure 6b, the following formulae can be obtained.

$$C_1 = \varepsilon_0\varepsilon_{eff2}\pi R^2/d_1 \tag{5}$$

$$C_2 = 4\varepsilon_0\varepsilon_{eff1}\pi R/\log(\sec(\pi(p-2R)/2p)) \tag{6}$$

$$L = -\mu_0 d_1(4+\ln(p-2r))/2\pi \tag{7}$$

$$f_0 = 1/2\pi\sqrt{LC_1C_2/(C_1+C_2)} \tag{8}$$

$$FBW = \sqrt{L(C_1+C_2)/C_1C_2}/\eta \tag{9}$$

where the vacuum dielectric constant is $\varepsilon_0 = 8.854 \times 10^{-12}$ F/m, vacuum permeability is $\mu_0 = 4\pi \times 10^{-7}$ H/m, effective impedance is η, FBW is the fractional bandwidth around the center frequency f_0, p is the permutation period between the TSV-based EBGs, r is the radius of the column, R is the radius of the disc, d_1 is the height of the column, d_1 is the thickness of the dielectric plate above the disc, C_1 is the capacitance of the disc with the top PEC, C_2 is the gap capacitance between the two discs, and L is the inductance of the copper column.

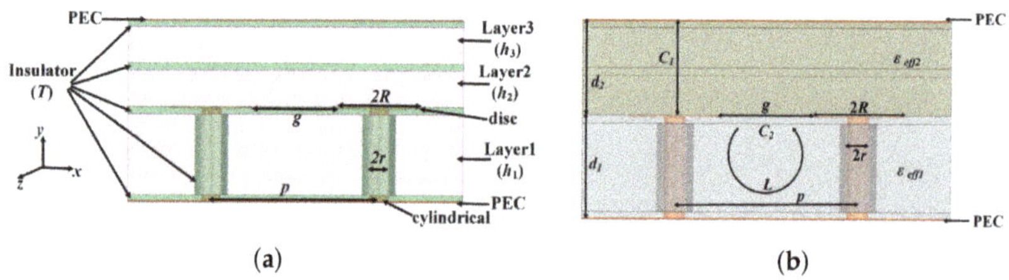

Figure 6. Side view of TSV-based EBGs: (**a**) physical structure and (**b**) equivalent physical structure.

Firstly, it is assumed that the height of the copper column is d_1 = 65 μm, the thickness of the medium plate above the copper column is d_2 = 60 μm, the center frequency is f_0 = 0.19 THz, and the fractional bandwidth is FBW = 65%. Additionally, according to formulae (5)–(9), we obtain R = 47.8 μm and r = 6.6 μm. Lastly, for convenience of the design, set R = 50 μm and r = 7 μm.

According to [14], it can be known that $Cox(f)$ tends to 0 at the frequency infinity. This work considers the THz band, which suggests $Cox(f)$ = 0.03 pF, and thus according to formula (1), T > 3.5 μm is obtained. Therefore, set T = 5 μm.

The designed TSV-based EBG structure unit is shown in Figure 7. The upper surface of the dielectric substrate (Layer 1) is coated with a layer of SiO_2. A metal disc with a diameter of 2R is printed on the upper surface of the insulator. And the lower surface of Layer 1 is also coated with a layer of SiO_2. A copper column with a diameter of 2r is inserted into the substrate, and a layer of SiO_2 with a thickness of T is wrapped outside the copper pillar. The upper surface of the middle dielectric substrate (Layer 2, Si) is covered with a layer of SiO_2. The upper surface of the upper dielectric substrate (Layer 3, Si) is also coated with a layer of SiO_2, and on the upper surface, there is a copper-clad metal layer. The dimensions of the TSV-based EBG are listed in Table 1.

Table 1. Dimensions of TSV-based EBG (units: μm).

T	p	r	R	h_1	h_2	h_3
5	120	7	50	55	25	25

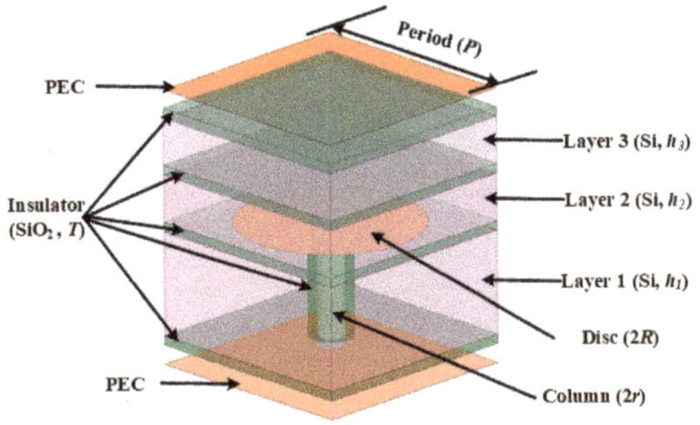

Figure 7. TSV-based EBG cell structure.

The electromagnetic simulation of the TSV-based EBG unit was conducted using CST Microwave Studio v2020 software. The dispersion is illustrated in Figure 8, revealing a central frequency of approximately 0.2 THz and a bandwidth of about 62%, aligning well with the design specifications.

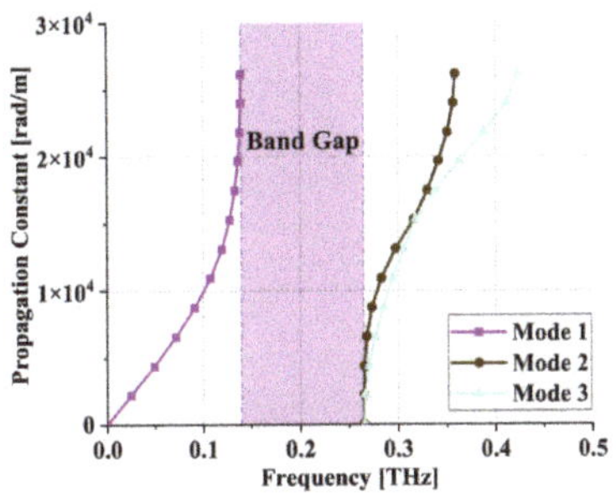

Figure 8. Dispersion of TSV-based EBG.

The SSIGW is composed of stripline and TSV-based EBG array, as illustrated in Figure 9. Further optimization using the High-Frequency Structure simulator (HFSS) determines a stripline width of 20 µm. The metal stripline is etched onto the upper surface of layer SiO_2 on the middle dielectric substrate (Layer 2, Si). The TSV-based EBG structure ensures that signal energy is confined to the stripline, minimizing energy diffusion leakage on both sides of the stripline.

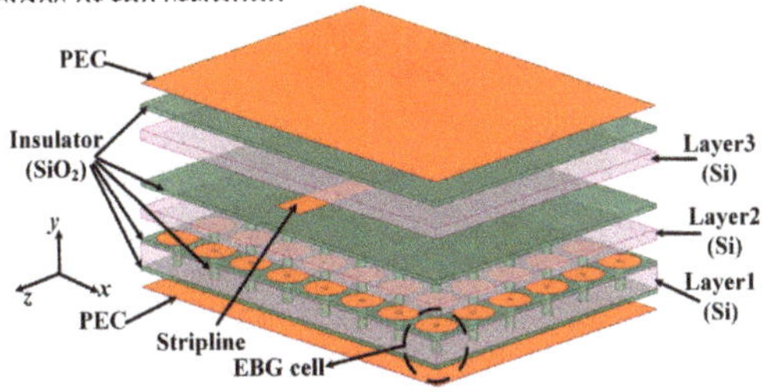

Figure 9. Physical structure of SSIGW.

CST Microwave Studio software is used to analyze the dispersion diagram of the SSIGW, as shown in Figure 10a. It is seen that the electromagnetic bandgap is 0.139–0.259 THz. The S parameters, $|S_{11}|$ and $|S_{21}|$, of the SSIGW are shown in Figure 10b. The −10 dB bandwidth of the SSIGW starts from 0.126 THz.

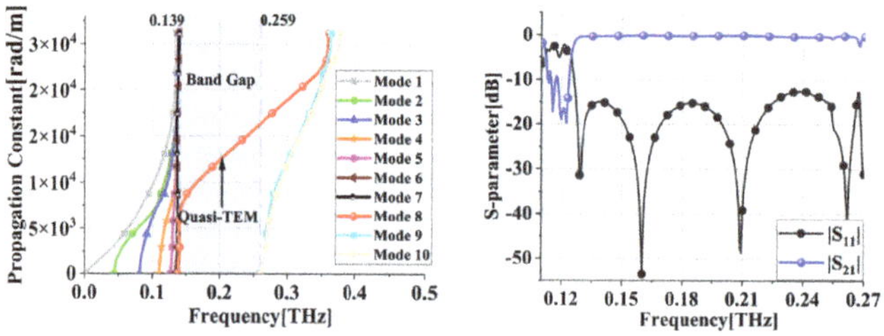

(a) (b)

Figure 10. Simulation results of the SSIGW: (**a**) dispersion and (**b**) $|S_{11}|$ and $|S_{21}|$.

3.2. Design and Analysis of Ultra-Wideband Bandpass Filter

The subsection improves a UWB bandpass filter based on SSIGW in [15], and its structure is shown in Figure 11a. A non-equal-impedance branch-and-loop resonator is proposed to solve the problem of excessive in-band return loss in conventional equal-impedance branch-and-loop resonators. In Figure 11b, the schematic depicts the UWB filter utilizing a circular ring resonator loaded with a pair of quarter wavelength open circuit stubs connected to the right and top center of the ring. The input and output ports are positioned at two orthogonal locations, specifically at the bottom and left center of the ring, and are directly linked to the ring. The equivalent transmission line model is illustrated in Figure 11c. The circumference of the ring is one wavelength, and the characteristic impedance is Z_1, while the electrical length of the open circuit stub is θ_2 and the characteristic impedance is Z_2.

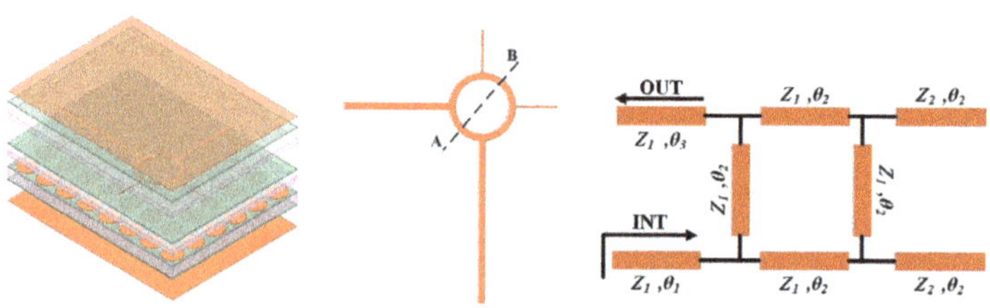

(a) (b) (c)

Figure 11. UWB bandpass filter: (**a**) 3D view; (**b**) 2D view of the resonator; and (**c**) equivalent transmission line model (Z_1 = 50 Ω, Z_2 = 100 Ω, θ_1 = 180°, θ_2 = 90°, θ_3 = 270°).

Under weak coupling [25,26], the plane of symmetry AB shown in Figure 11b can be considered as an ideal magnetic or electrical wall, respectively. The equivalent even/odd mode transmission line model of the ring resonator used to determine the resonant frequency of the filter is shown in Figure 12a and Figure 12b, respectively.

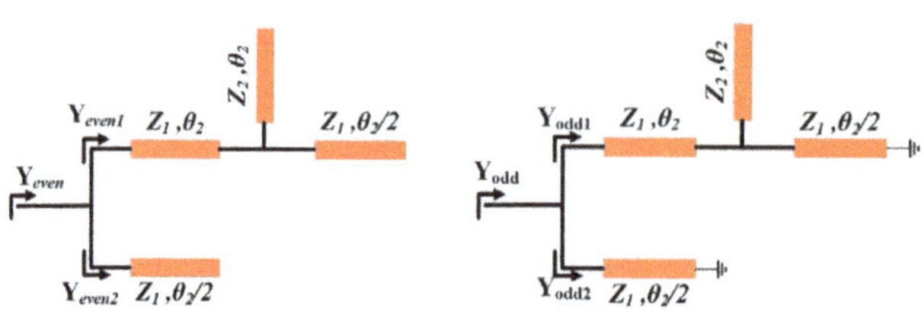

(a) (b)

Figure 12. Even and odd mode analysis: (**a**) even mode transmission line model; (**b**) odd mode transmission line model.

According to the ring resonator theory, it can be concluded that the ring resonator resonates when its input admittance is zero [25]. Therefore, the resonance conditions are shown as follows:

$$Y_{even} = Y_{even1} + Y_{even2} = 0 \tag{10}$$

$$Y_{odd} = Y_{odd1} + Y_{odd2} = 0 \tag{11}$$

In this subsection, the odd–even mode method is used to calculate the resonance frequency of a circular ring resonator loaded with an open circuit stub. The even mode equivalent transmission line model when a magnetic wall is applied to the symmetric plane AB of the ring resonator is shown in Figure 12a. The analysis is simplified by neglecting all effects of transmission line discontinuities within the ring resonator depicted in Figure 11b. The normalized even mode input admittances Y_{even1} and Y_{even2} of the upper and lower arms are, respectively, expressed by the following formula:

$$Y_{even1} = j\frac{-m^3 - Knm^2 + 3m + Kn}{-3m^2 - 2Knm + 1} \tag{12}$$

$$Y_{even2} = jm \tag{13}$$

where $m = \tan(\theta_2/2)$, $n = \tan\theta_2 = 2m/(1-m^2)$, $K = Z_1/Z_2$.

Thus, the even mode resonance condition (10) can be obtained by the following, equation:

$$Y_{even} = -j\frac{4m^3 + 3Knm^2 - 4m - Kn}{-3m^2 - 2Knm + 1} = 0 \tag{14}$$

Odd mode analysis is akin to even mode. In Figure 12b, the equivalent transmission line model represents a half-ring resonator with an electrical wall applied along AB. The normalized odd mode input admittances Y_{odd1} and Y_{odd2} of the upper and lower arms can be deduced as follows:

$$Y_{odd1} = -j\frac{Knm^3 - 3m^2 - Kmn + 1}{-m^3 - 2Knm^2 + 3m} \tag{15}$$

$$Y_{odd2} = 1/jm \tag{16}$$

Therefore, the odd mode resonance condition (11) can be obtained by the following equation:

$$Y_{odd} = j\frac{Knm^3 - 4m^2 - 3Knm + 4}{m^3 + 2Knm^2 - 3m} = 0 \tag{17}$$

If the loop is assumed to be lossless, the frequency at which the transmission zero is located can be obtained as shown in [26], by solving the following:

$$Y_{even} = Y_{odd} \tag{18}$$

By substituting (14) and (17) into (18), the following polynomial equation can be obtained.

$$m^6 + 2Knm^5 + \left(K^2n^2 - 1\right)m^4 + \left(K^2n^2 - 1\right)m^2 - 2Knm + 1 = 0 \tag{19}$$

The resonant frequency can be calculated by the following formula:

$$f_r = \frac{2\pi c}{\theta_2 \lambda_g \sqrt{\varepsilon_{eff_2}}} \tag{20}$$

where c is the speed of light in free space, and λ_g is the waveguide wavelength. This excerpt uses $K = 1/2$, $\lambda_g = 0.61$ mm, $\varepsilon_{eff_2} = 10.583$.

Table 2 shows the comparison between the calculated and simulated resonance frequencies. It can be seen that the calculated values agree with the simulations. The causes of errors include loss of loops, discontinuity, and so on. In addition, the calculation has only two transmission zeros, while the simulation has four. The additional transmission zeros are generated by SSIGW itself, as can be seen from Figure 10b.

Table 2. Comparison of calculated and simulated resonant frequencies (units: THz).

Resonant Frequency	Calculated	Simulated
$TZ1$	0.128	0.130
$TZ2$	–	0.133
$TZ3$	0.224	0.234
$TZ4$	–	0.241
$TP1$	0.151	0.144
$TP2$	0.191	0.191
$TP3$	0.220	0.224

Simulated $|S_{11}|$ and $|S_{21}|$ are shown in Figure 13a. As can be seen from Figure 13a, its passband is 0.142–0.226 THz, and its fractional bandwidth (FBW) is 45.34% around the center frequency of 0.184 THz. The insertion loss is lower than 0.71 dB, and the out-of-band suppression is higher than 20 dB. Simulated group delay is shown in Figure 13b. As can be seen from Figure 13b, its in-band group delay is 0.037 ± 0.023 ns, very flat over the bandwidth.

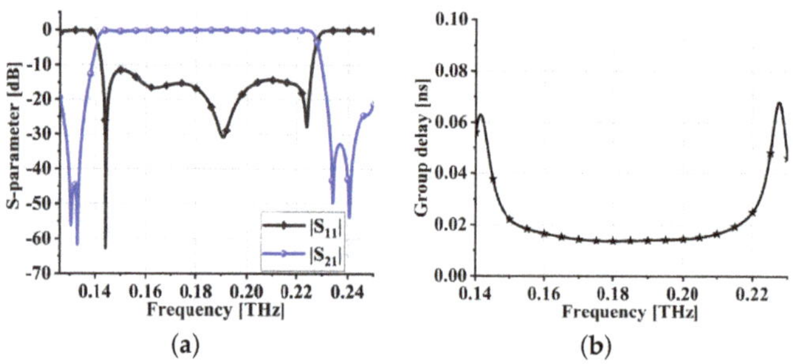

Figure 13. Simulation results of UWB bandpass filter: (**a**) S parameters and (**b**) in-band group delay.

3.3. Design and Analysis of Meta-Surface Wideband Bandstop Filter

This subsection introduces a novel meta-surface bandstop filter, depicted in Figure 14a. The stopband functionality is achieved by adding SRRs on both sides of the transmission line. Adjusting the dimensions of the SRRs allows for the realization of stopbands in different frequency ranges.

The planar view and corresponding equivalent circuit are depicted in Figure 14b and Figure 14c, respectively. Figure 14d illustrates the circuit unit. To elucidate the operational principles, a detailed analysis of the equivalent circuit unit follows. Empirical formulas (21)–(24) are derived from both [27,28] and the SSIGW physical model presented in this paper. The calculation of the dimensions of meta-surface bandstop filter 1 yields L_{SRR} = 0.0272 nH, C_{SRR} = 0.033 pF. Following optimization using the Advanced Design System (ADS), the parameters are confirmed to be L_{SRR} = 0.0272 nH, C_{SRR} = 0.035 pF, and the S-parameter is displayed in Figure 15a. From Figure 15b, it can be observed that the resonant points from theoretical calculations and simulations coincide at 0.162 THz.

$$L_{SRR} = \frac{\mu_r \mu_0 \ln(1 + 32(8h_3 + T)^2(1 + \sqrt{1 + \pi w_{M1}^2/(8h_3 + T)/w_{M1}^2})}{100000\pi} \tag{21}$$

$$C_{SRR} = \frac{\varepsilon_{eff2}\varepsilon_0 M}{9000} \tag{22}$$

$$M = \frac{2\ln(2(1+\sqrt{K})/(1-\sqrt{K}))}{\pi} \tag{23}$$

$$K = \sqrt{1 - \frac{1}{(1+2w_{M1}/S_{M1})^2}} \tag{24}$$

in which $\varepsilon_{eff2} = 10.583$ represents the relative permittivity, $\mu_r = 1$ is the relative magnetic permeability, w_{M1} = 8 μm is the metal line width of the SRR, and S_{M1} = 8 μm is the gap distance between SRR openings.

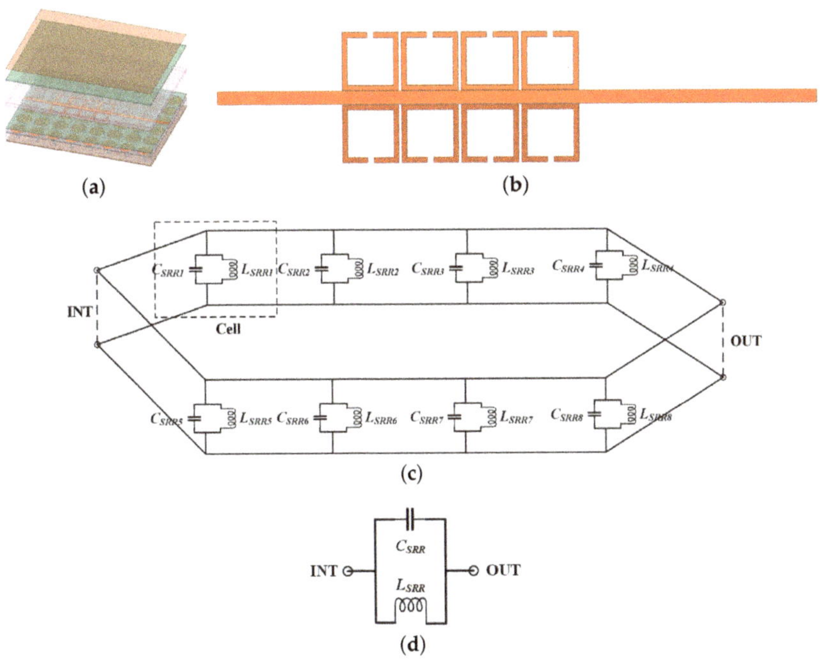

Figure 14. Meta-surface bandstop filter: (**a**) 3D view; (**b**) 2D view of the filter resonator; (**c**) the equivalent circuit; and (**d**) the circuit unit.

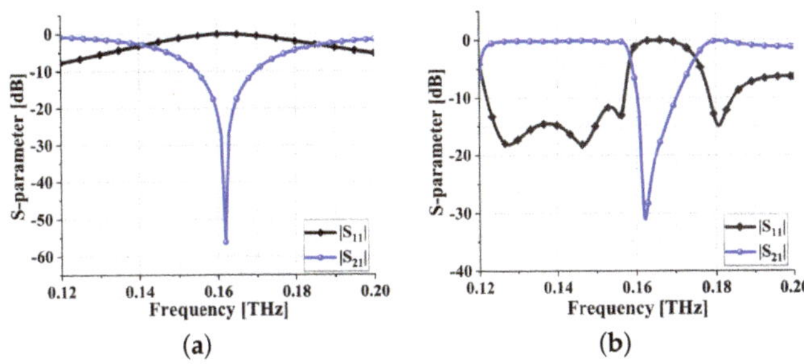

Figure 15. S parameters of filter: (**a**) equivalent circuit; (**b**) meta-surface bandstop filter 1.

As can be seen from Figure 16a, inside the stopband (@0.165 THz), energy is absorbed by meta-surface losses, especially on the first unit. As can be seen from Figure 16b,c, outside the stopband (@0.14 THz and @0.18 THz), energy concentrates on the transmission line and is propagated to the output end. This further validates the accuracy of the theoretical analysis mentioned above.

In order to broaden the bandwidth, this paper designed meta-surface bandstop filter 2 with adjacent stopband ranges. The S-parameters, as shown in Figure 17a, reveal a stopband range of 0.168–0.18 THz. Overlaying the two configurations resulted in a meta-surface filter with a wide stopband, as depicted in Figure 17b, now exhibiting a stopband range of 0.16–0.181 THz. This configuration effectively satisfies the criteria for broad stopband performance.

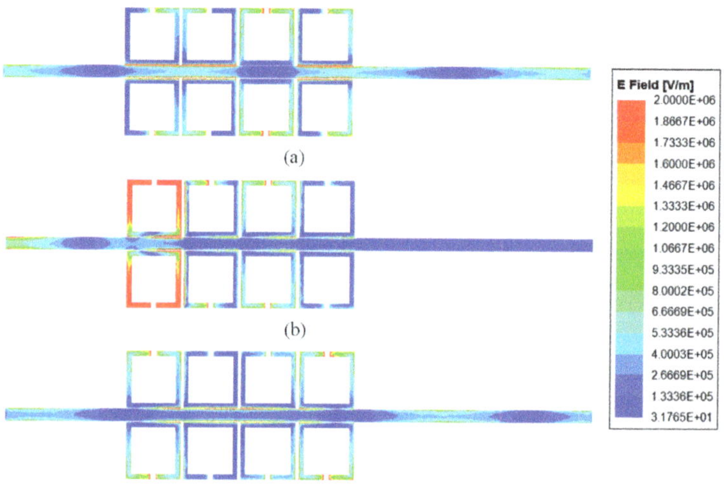

Figure 16. Electric field profile of meta-surface bandstop filter 1: (**a**) outside the stopband (@0.14 THz); (**b**) in-band (@0.165 THz); and (**c**) outside the stopband (@0.18 THz).

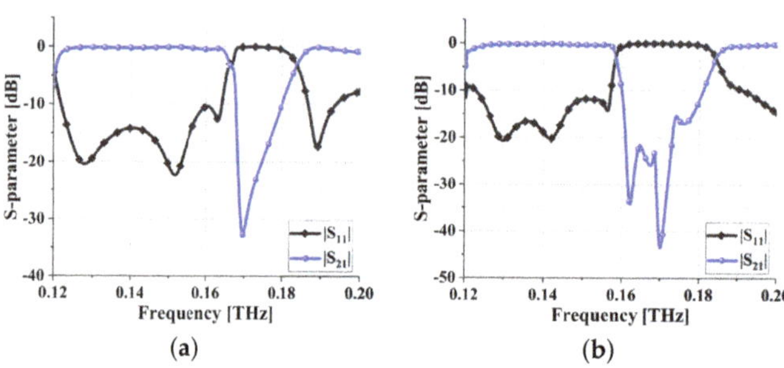

Figure 17. S parameters of filters: (**a**) meta-surface bandstop filter 2; and (**b**) wide stopband meta-surface bandstop filter.

3.4. Results and Comparisons

The parameters and their corresponding values for the filter proposed in this paper, after the final optimization, are presented in Table 3. The specific physical description of these parameters is illustrated in Figure 18. The simulation results of the proposed THz filter are shown in Figure 19. From Figure 19a, it can be observed that the center frequency is 0.151 THz, the FBW is 6.9%, the insertion loss is less than 0.68 dB, and both the upper and lower stopband widths exceed 20 GHz. Simulated group delay is shown in Figure 19b. As can be seen from Figure 19b, the in-band group delay is 0.119 ± 0.048 ns, again very flat.

Table 3. Dimensions of meta-surface THz bandpass filter (units: μm).

Parameters	Values	Parameters	Values	Parameters	Values
L_1	1738	L_2	352	L_3	125
R_1	98	w_1	20	w_2	7
L_{M1}	90	S_{M1}	8	w_{M1}	8
L_{M2}	83	S_{M2}	3	w_{M2}	8
G_{M11}	9	G_{M12}	5	G_1	200
G_{M21}	9	G_{M22}	8	G_1	50

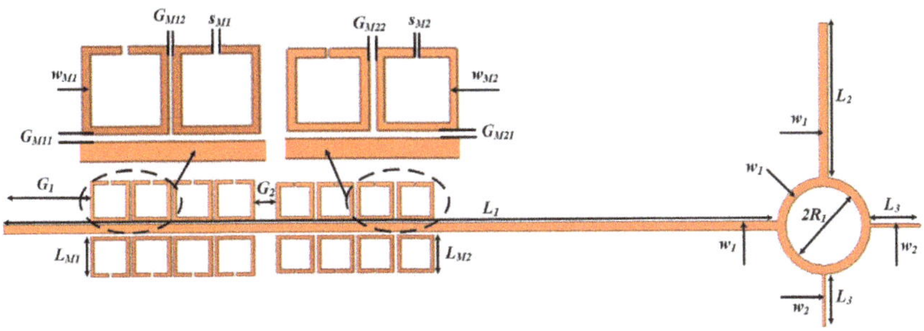

Figure 18. Parameters of meta-surface THz bandpass filter.

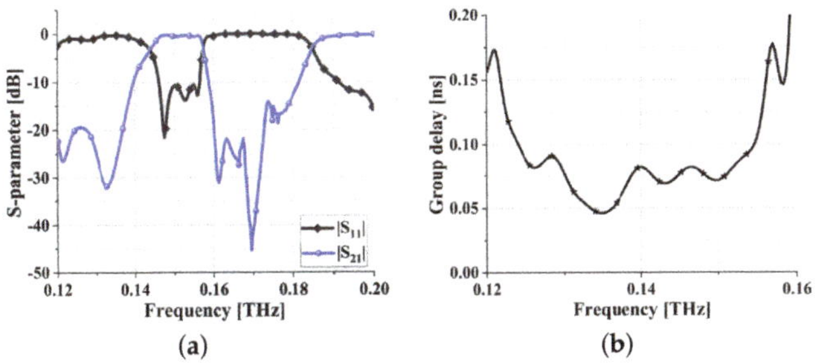

Figure 19. Simulation results of meta-surface THz bandpass filter: (**a**) S parameters and (**b**) in-band group delay.

Due to the influence of processing errors on the performance of filters, and considering that the processing accuracy of TSV can reach the nanometer level, this subsection simulates the filter with an error of ±0.1 μm. The simulation results are shown in Figure 20, where Figure 20a represents the $|S_{11}|$ with a ±0.1 μm error, and Figure 20b represents the $|S_{21}|$ with a ±0.1 μm error. It can be observed that the actual processing errors of the filter designed in this study have almost no impact, demonstrating high stability. Furthermore, the operating temperature can also contribute to changes in the performance of the filter. In this study, TSV technology is employed, and the impact of the operating temperature on the filter is determined by the TSV operating temperature. According to [29,30], it can be concluded that the stable operating temperature range for the filter proposed in this paper is from −40 °C to 120 °C.

To further demonstrate the universality of the theoretical framework proposed in this article, this section presents the filtering results in another frequency band, as shown in Figure 21. From Figure 21a, it can be observed that the center frequency is 2.215 THz, the FBW is 6.1%, the insertion loss is less than 1.29 dB, and both the upper and lower stopband widths exceed 50 GHz. Simulated group delay is shown in Figure 21b. As can be seen from Figure 21b, the in-band group delay is 0.007 ± 0.004 ns, again very flat.

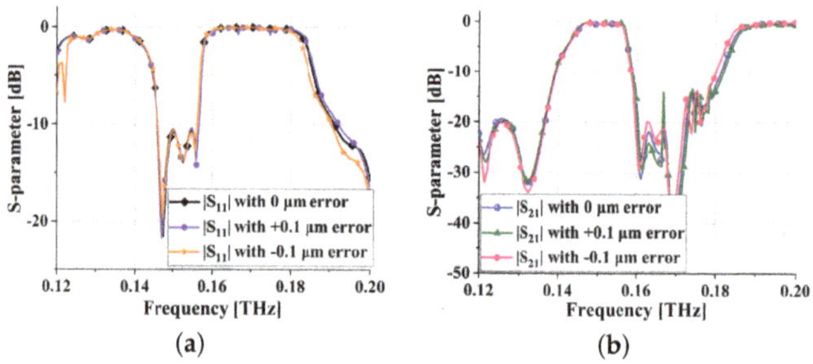

Figure 20. Error simulation results of meta-surface THz bandpass filter: (**a**) $|S_{11}|$ and (**b**) $|S_{21}|$.

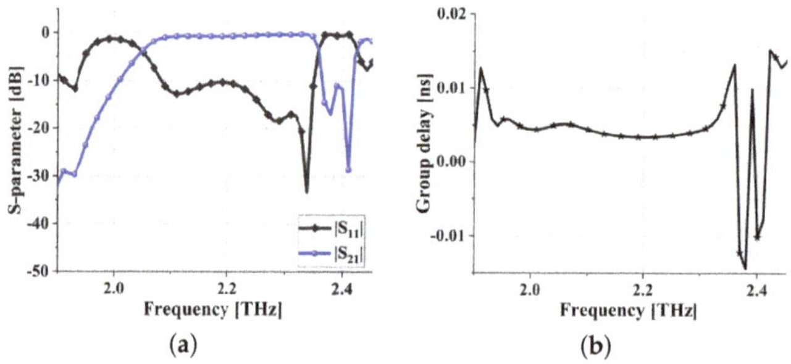

Figure 21. Simulation results of another meta-surface THz bandpass filter: (**a**) S parameters and (**b**) in-band group delay.

A comparison of the above designed filters with the reported six related filters is performed in Table 4. Note that the SSIGW filter showcases ultra-low latency and exceptional filtering characteristics. In comparison to the existing literature [3,4], the work demonstrates smaller insertion losses and is more easily integrable than filters designed using SIW, as SSIGW supports TEM mode. Unlike previous studies, the work excels in achieving lower insertion losses than filters designed using other techniques. Additionally, the proposed filter exhibits a 21-fold reduction in group delay compared to [4], making it more suitable for low-latency communication in the THz range.

Table 4. Comparison with related THz filters.

Filters	Type	Method	Center Frequency [THz]	FBW [%]	Insertion Loss [dB]	Group Delay [ns]
[1]	Meta-material	Sim.	7	22.85	- -	3.67
[2]	Meta-material	Sim.	0.125	65	- -	- -
[3]	SIW	Meas.	0.331	15.4	1.5	- -
[4]	SIW	Sim.	0.160	12.5	1.5	2.5 ± 1.5
[13]	Hairpin	Sim.	0.500	16	1.5	- -
[31]	CNC milling	Meas.	0.292	14.64	3	- -
[32]	Meta-material	Sim.	1	18	3	- -
[33]	Meta-material	Meas.	0.6	- -	>3	- -
[34]	Meta-material	Sim.	2.37	59	3	- -
This work 1	SSIGW	Sim.	0.151	6.9	0.68	0.119 ± 0.048
This work 2	SSIGW	Sim.	2.215	6.1	1.29	0.007 ± 0.004

4. Conclusions

THz filters currently face challenges such as significant losses, integration difficulties, and complexities in designing filters with arbitrary bandwidths. This paper proposes a meta-surface-based approach for designing arbitrary bandwidth SSIGW filters, effectively addressing the aforementioned challenges. The method achieves bandwidth control by incorporating a meta-surface-based bandstop filter into a UWB filter and controlling the operating frequency range of the meta-surface bandstop filter. Additionally, the utilization of SSIGW enables functionalities such as low loss and easy integration. To validate the effectiveness and superiority of the proposed method, a THz filter is designed in this study. In the UWB filtering section, non-equal-impedance branch-and-loop resonator is employed, and the analysis is conducted using transmission line theory, with the theoretical and simulated results aligning closely. The bandstop filtering section utilizes SRR meta-surfaces, analyzed through an equivalent circuit analysis, with the theoretical and simulated results aligning closely. Adjusting the operating frequency range of the SRR meta-surface allows for different bandwidths, while increasing the number of SRRs enables a wider stopband. The presented filter exhibits an insertion loss below 0.68 dB, with a stopband width exceeding 20 GHz and a group delay of 0.119 ± 0.048 ns. Compared to existing THz filters, it exhibits advantages such as lower loss, lower group delay, and controllable arbitrary bandwidth. Therefore, it is expected to become a candidate for THz wireless communication systems.

Author Contributions: Conceptualization, D.H. and L.W.; methodology, D.H.; software, D.H.; validation, L.W. and Q.L.; formal analysis, D.H.; investigation, D.H., L.W. and Q.L.; resources, Z.L., X.X., H.C. and Y.L.; data curation, D.H.; writing—original draft preparation, D.H.; writing—review and editing, L.W., Q.L., Z.L., X.X., H.C. and Y.L.; visualization, D.H.; supervision, Z.L., X.X., H.C. and Y.L.; project administration, Z.L.; funding acquisition, Z.L., X.X. and H.C. All authors have read and agreed to the published version of the manuscript.

Funding: This research is supported by the National Key Research and Development Program of China, grant number 2021YFB2900401, the Key Research and Development Program of Shenzhen, grant number ZDSYS20210623091807023, and the Major Key Project of PCL Department of Broadband Communication, grant number PCL2023AS1-1.

Institutional Review Board Statement: Not applicable.

Informed Consent Statement: Not applicable.

Data Availability Statement: Data are contained within the article.

Conflicts of Interest: The authors declare no conflicts of interest.

References

1. Caroline, B.E.; Sagadevan, K.; Danasegaran, S.K.; Kumar, S. Characterization of a Pentagonal CSRR Bandpass Filter for Terahertz Applications. *J. Electron. Mater.* **2022**, *51*, 5405–5416. [CrossRef]
2. Ruan, J.F.; Lan, F.; Wang, L.; Ji, S.W. Ultra-wideband THz metamaterial filter with steep cut-off. *J. Electromagn. Waves Appl.* **2021**, *35*, 431–440. [CrossRef]
3. Wang, F.; Pavlidis, V.F.; Yu, N. Miniaturized SIW bandpass filter based on TSV technology for THz applications. *IEEE Trans. Terahertz Sci. Technol.* **2020**, *10*, 423–426. [CrossRef]
4. Liu, X.; Zhu, Z.; Liu, Y.; Lu, Q.; Yin, X.; Yang, Y. Wideband substrate integrated waveguide bandpass filter based on 3-D ICs. *IEEE Trans. Compon. Packag. Manuf. Technol.* **2019**, *9*, 728–735. [CrossRef]
5. Attia, H.; Sorkherizi, M.S.; Kishk, A.A. 60 GHz slot antenna array based on ridge gap waveguide technology enhanced with dielectric superstrate. In Proceedings of the 2015 9th European Conference on Antennas and Propagation (EuCAP), Lisbon, Portugal, 13–17 April 2015; pp. 1–4.
6. Li, T.; Chen, Z.N. Wideband Sidelobe-Level Reduced *Ka* -Band Metasurface Antenna Array Fed by Substrate-Integrated Gap Waveguide Using Characteristic Mode Analysis. *IEEE Trans. Antennas Propag.* **2020**, *68*, 1356–1365. [CrossRef]
7. Abbas, M.A.; Cengiz, M.F.; Allam, A.M.M.A.; Fawzy, D.E.; Elhennawy, H.M.; Sree, M.F.A. A Novel Circular Reconfigurable Metasurface-based Compact UWB Hybrid Coupler for Ku-band Applications. *IEEE Access* **2022**, *10*, 129781–129790. [CrossRef]
8. Sifat, S.M.; Shams, S.I.; Kishk, A.A. Millimeter-Wave Ferrite Circulator Integrated on Gap Waveguide Technology. *IEEE Microw. Wirel. Technol. Lett.* **2023**, *33*, 1151–1154. [CrossRef]
9. Yin, X.; Zhu, Z.; Yang, Y.; Ding, R. Effectiveness of $p+$ layer in mitigating substrate noise induced by through-silicon via for microwave applications. *IEEE Microw. Wirel. Compon. Lett.* **2016**, *26*, 687–689. [CrossRef]
10. Qian, L.; Xia, Y.; He, X.; Qian, K.; Wang, J. Electrical modeling and characterization of silicon-core coaxial through-silicon vias in 3-D integration. *IEEE Trans. Compon. Packag. Manuf. Technol.* **2018**, *8*, 1336–1343. [CrossRef]

11. Yin, X.; Zhu, Z.; Liu, Y.; Lu, Q.; Liu, X.; Yang, Y. Ultra-compact TSV-based L-C low-pass filter with stopband up to 40 GHz for microwave application. *IEEE Trans. Microw. Theory Tech.* **2019**, *67*, 738–745. [CrossRef]
12. Hu, S.; Wang, L.; Xiong, Y.-Z.; Lim, T.G.; Zhang, B.; Shi, J.; Yuan, X. TSV technology for millimeter-wave and terahertz design and applications. *IEEE Trans. Components Packag. Manuf. Technol.* **2011**, *1*, 260–267. [CrossRef]
13. Wang, F.; Ke, L.; Yin, X.; Yu, N.; Yang, Y. Compact TSV-based hairpin bandpass filter for thz applications. *IEEE Access* **2021**, *9*, 132078–132083. [CrossRef]
14. Li, W.; Liu, Z.; Qian, W.; Wang, Z.; Wang, W.; Zhao, Y.; Zhang, X. Modeling of the RF coaxial TSV configuration inside the silicon interposer with embedded cooling cavity. *IEEE Trans. Compon. Packag. Manuf. Technol.* **2022**, *12*, 3–10. [CrossRef]
15. Hou, D.; Lin, S.; Shen, D.; Yuan, H.; Wang, B.; Zhang, X. An Integrated Substrate Gap Waveguide Filter Metalens Antenna for 6G THz Band. In Proceedings of the 2022 IEEE MTT-S International Microwave Workshop Series on Advanced Materials and Processes for RF and THz Applications (IMWS-AMP), Guangzhou, China, 13–15 November 2022; pp. 1–3.
16. Minnis, B.J. Classes of Sub-Miniature Microwave Printed Circuit Filters with Arbitrary Passband and Stopband Widths. *IEEE Trans. Microw. Theory Tech.* **1982**, *30*, 1893–1900. [CrossRef]
17. Ma, Y.H.; Yuan, Y.; Yuan, W.T.; Wu, W.W.; Yuan, N.C. A Novel Method to Design Stub-Loaded Microstrip Filters with Arbitrary Passband Based on the Reflection Theory and Monte Carlo Method. *Radio Eng.* **2019**, *28*, 136–146. [CrossRef]
18. Luo, S.; Mei, P.; Zhang, Y.; Pedersen, G.F.; Zhang, S. Decoupling of Dual-Polarized Antenna Arrays Using Non-Resonant Metasurface. *Sensors* **2023**, *23*, 152. [CrossRef]
19. Khan, M.S.; Khan, S.; Khan, O.; Aqeel, S.; Gohar, N.; Dalarsson, M. Mutual Coupling Reduction in MIMO DRA through Metamaterials. *Sensors* **2023**, *23*, 7720. [CrossRef]
20. Tataria, H.; Shafi, M.; Molisch, A.F.; Dohler, M.; Sjöland, H.; Tufvesson, F. 6G Wireless Systems: Vision, Requirements, Challenges, Insights, and Opportunities. *Proc. IEEE* **2021**, *109*, 1166–1199. [CrossRef]
21. Oliver, W.C.; Pharr, G.M. An improved technique for determining hardness and elastic modulus using load and displacement sensing indentation experiments. *J. Mater. Res.* **1992**, *7*, 1564–1583. [CrossRef]
22. John, H.L. Through-Silicon Vias: Mechanical, Thermal, and Electrical Behaviors. In *Through-Silicon vias for 3D Integration (Reading Version)*; Publishing House: Beijing, China, 2014; pp. 157–203.
23. Maslovski, S.I.; Morgado, T.A.; Silveirinha, M.G.; Kaipa, C.S.R.; Yakovlev, A.B. Generalized additional boundary conditions for wire media. *New J. Phys.* **2010**, *12*, 113047. [CrossRef]
24. Yang, F.; Rahmat-Samii, Y. Microstrip antennas integrated with electromagnetic band-gap (EBG) structures: A low mutual coupling design for array applications. *IEEE Trans. Antennas Propag.* **2003**, *51*, 2936–2946. [CrossRef]
25. Sun, S. A Dual-Band Bandpass Filter Using a Single Dual-Mode Ring Resonator. *IEEE Microw. Wirel. Components Lett.* **2011**, *21*, 298–300. [CrossRef]
26. Matsuo, M.; Yabuki, H.; Makimoto, M. Dual-mode stepped-impedance ring resonator for bandpass filter applications. *IEEE Trans. Microw. Theory Tech.* **2001**, *49*, 1235–1240. [CrossRef]
27. Holloway, C.L.; Hufford, G.A. Internal inductance and conductor loss associated with the ground plane of a microstrip line. *IEEE Trans. Electromagn. Compat.* **1997**, *39*, 73–78. [CrossRef]
28. Stellari, F.; Lacaita, A.L. New formulas of interconnect capacitances based on results of conformal mapping method. *IEEE Trans. Electron Devices* **2000**, *47*, 222–231. [CrossRef]
29. Katti, G.; Stucchi, M.; Velenis, D.; Soree, B.; Meyer, K.D.; Dehaene, W. Temperature-Dependent Modeling and Characterization of Through-Silicon Via Capacitance. *IEEE Electron Device Lett.* **2011**, *32*, 563–565. [CrossRef]
30. Chan, Y.C.; Basu, N.; Chen, T.-W.; Tsai, Y.-T.; Lin, H.-Y.; Chen, S.-C.; Lee, M.-H.; Liao, M.-H. The Analysis of Multiwall Carbon Nanotubes as Through Silicon Via by Equivalent Circuit Model at Different Operating Temperatures in Multilayers Stacking Scheme. *IEEE Trans. Electron Devices* **2023**, *70*, 3360–3364. [CrossRef]
31. Lin, C.H.; Cheng, Y.H. A THz Waveguide Bandpass Filter Design Using an Artificial Neural Network. *Micromachines* **2022**, *13*, 841. [CrossRef]
32. Chang, C.C.; Huang, L.; Nogan, J.; Chen, H.T. Invited Article: Narrowband terahertz bandpass filters employing stacked bilayer metasurface antireflection structures. *APL Photonics* **2018**, *3*, 051602. [CrossRef]
33. Huang, Y.; He, Q.C.; Zhang, D.P.; Kanamori, Y. Switchable band-pass filter for terahertz waves using VO_2-based metamaterial integrated with silicon substrate. *Opt. Rev.* **2021**, *28*, 92–98. [CrossRef]
34. Niharika, N.; Singh, S.; Kumar, P. Multifunctional metasurface based bandstop and bandpass filters for terahertz radiation. *Optik* **2022**, *253*, 168551. [CrossRef]

sensors

MDPI

Article

Design and Performance Analysis of Compact Printed Ridge Gap Waveguide Phase Shifters for Millimeter-Wave Systems

Moath Alathbah [1], Mohamed S. El-Gendy [2,*], Mahmoud Gadelrab [3] and Mohamed Mamdouh M. Ali [4]

[1] Department of Electrical Engineering, College of Engineering, King Saud University, Riyadh 11451, Saudi Arabia
[2] Microstrip Department, Electronics Research Institute, Cairo 11843, Egypt
[3] R&D Department, Scientific Microwave Corporation, Montreal, QC H4S 1C1, Canada
[4] Department of Electrical Engineering, Faculty of Engineering, Assiut University, Assiut 71516, Egypt
* Correspondence: mohamadalgendy2004@gmail.com

Abstract: This paper introduces compact Printed Ridge Gap Waveguide (PRGW) phase shifters tailored for millimeter-wave applications, with a focus on achieving wide operating bandwidth, and improved matching and phase balance compared to single-layer technology. This study proposes a unique approach to achieve the required phase shift in PRGW technology, which has not been previously explored. This study also introduces a novel analytical approach to calculate the cutoff frequency and propagation constant of the PRGW structure, a method not previously addressed. Furthermore, the utilization of multi-layer PRGW technology enables the realization of multi-layer beamforming networks without crossing, thereby supporting wideband operation in a compact size. The proposed design procedure enables the realization of various phase shift values ranging from 0° to 135° over a broad frequency bandwidth centered at 30 GHz. A 45-degree phase shifter is fabricated and tested, demonstrating a 10 GHz bandwidth (approximately 33% fractional bandwidth) from 25 GHz to 35 GHz. Throughout the operating bandwidth, the phase balance remains within $45 \pm 5°$, with a deep matching level of −20 dB. The proposed phase shifter exhibits desirable characteristics, such as compactness, low loss, and low dispersion, making it a suitable choice for millimeter-wave applications, including beyond 5G (B5G) and 6G wireless communications.

Keywords: cutoff frequency; effective permittivity; phase shifter; printed ridge gap waveguide (PRGW); propagation constant

1. Introduction

The rapid growth of millimeter-wave applications, driven by the increasing demand for high-speed wireless communication systems, such as fifth-generation (5G) networks, has sparked significant interest in the development of compact and efficient phase shifters. Phase shifters play a critical role in beamforming, antenna steering, and signal processing, enabling enhanced data rates, improved coverage, and increased network capacity. While 5G continues to revolutionize the way in which we interact with the world, the evolution to 6G is already in the early applied research phase, emphasizing the need for advanced millimeter-wave technologies [1–3]. In this context, a substantial body of research has been dedicated to exploring novel techniques and technologies for achieving desirable phase shifter characteristics, particularly for applications such as beamforming networks, including Butler matrices. Compactness is crucial for realizing the full potential of phase shifters in these networks, as it enables the integration of multiple components in a limited space, facilitating the advancement of 6G technical possibilities and trends [4–6].

The current research landscape exposes a significant gap in advancing phase shifter technology using Printed Ridge Gap Waveguides (PRGWs) for millimeter-wave frequencies. While there has been some progress, it is worth noting that only one study [7] has managed to utilize Schiffman phase shifters effectively to construct a linear Butler matrix and implement a 1D beam scanning antenna array. However, this approach faced limitations due to its narrow operational bandwidth, restricted to 20%, and large size. The Schiffman technique, although effective, encounters challenges at mmWave frequencies because it relies on coupled line theory, requiring small gaps between lines that are difficult to fabricate accurately at higher frequencies, thus limiting its practicality. On the other hand, alternative techniques, such as time delay, used in PMC packaged or Inverted Microstrip Line Gap Waveguides, share similarities with PRGW but demand additional dielectric layers, leading to increased costs and complexity. Despite efforts to improve performance with matching stubs, these implementations still exhibit significant phase imbalances over a limited bandwidth, typically around 16% [8]. Recognizing the limitations of single-layer network structures in terms of compactness and bandwidth, there has been a shift towards multi-layer configurations in beamforming networks. Multi-layer setups offer the advantage of compacting beamforming networks while minimizing phase imbalance [9]. Moreover, various alternative mechanisms based on Metallic Ridge Gap Waveguides, such as ferrite-based, dielectric-filled, and glide symmetrical holes [10–12], have been explored. However, these structures often suffer from issues like large size, limited bandwidth, and poor phase imbalance. Additionally, implementing a multi-layer configuration requires a high-tolerance fabrication process, which is currently not cost-effective.

These collective challenges underscore the urgent need for further research and innovation in millimeter-wave phase shifter design. Specifically, there is a critical demand for the development of compact, wideband phase shifters with minimal phase imbalances, particularly for application in multi-layer beamforming networks. Addressing these challenges will be paramount for advancing the capabilities of future wireless communication systems, including the anticipated evolution to 6G technology. Building upon previous work, this paper presents a comprehensive investigation of compact ultra-wideband printed ridge gap waveguide phase shifters for millimeter-wave applications. The main focus of this study is to overcome the limitations of single-layer technology by adopting a multi-layer coupling technique between two resonant patches. This approach offers a wider operating bandwidth and improved matching level and phase balance compared to previous designs. Additionally, this study introduces a novel analytical method for calculating the cutoff frequency and propagation constant of the PRGW structure, providing a new way to characterize PRGW structures that has not been addressed before. A systematic design procedure is employed to achieve a range of phase shift values from 0° to 135° over a broad frequency bandwidth centered at 30 GHz. To validate the effectiveness of the proposed design approach, a 45° phase shifter is fabricated and experimentally evaluated. The measurement results demonstrate an impressive bandwidth of 10 GHz, covering frequencies from 25 GHz to 35 GHz, with a phase balance within ±5° throughout the operating bandwidth and a deep matching level of −25 dB. These findings highlight the superior characteristics of the proposed phase shifter, including its compactness, low loss, and low dispersion, positioning it as a promising candidate for millimeter-wave applications, particularly in B5G/6G wireless communications.

This paper is structured as follows: Section 2 presents a theoretical analysis of the PRGW structure, introducing a novel analytical approach to calculate the cutoff frequency and propagation constant, which is crucial for characterizing the PRGW structure. Section 3 focuses on the proposed PRGW phase shifter, detailing the theoretical analysis and design methodology, including the multi-layer PRGW phase shifter analysis, the proposed design, and the design methodology, along with simulation results. Section 4 presents the measurements and validation of the fabricated phase shifter, including comparisons with state-of-the-art designs to highlight its advantages. Finally, Section 5 concludes the paper by summarizing the findings and suggesting potential directions for future research.

2. Theoretical Analysis of PRGW Structure

In the field of electromagnetic engineering, while perfect electric conductors (PECs) are commonly found in nature, perfect magnetic conductors (PMCs) are engineered to exhibit properties resembling ideal magnetic surfaces within specific frequency bands [13–15]. These engineered PMC surfaces, also known as artificial magnetic conductors (AMCs) or band-gap surfaces, serve as essential components in various electromagnetic applications, facilitating precise control over wave propagation and manipulation of electromagnetic fields. One notable technology that exploits the capabilities of AMC surfaces is the Ridge Gap Waveguide (RGW) [16–20]. RGW configurations typically consist of two parallel plates, one featuring carefully designed periodic textures intended to control the propagation of electromagnetic waves. By purposefully arranging these periodic textured cells, RGW systems effectively confine electromagnetic energy within the waveguide structure, thereby preventing wave leakage and maintaining a quasi-TEM (transverse electromagnetic) mode within the air gap. This arrangement utilizes the unique properties of AMC surfaces, enabling RGW to establish a parallel-plate band-gap and facilitate precise control over electromagnetic waves in millimeter-wave and terahertz frequency regimes.

RGW technology encompasses various configurations, one of which is the Printed Ridge Gap Waveguide (PRGW). The development of PRGW represents a significant advancement in electromagnetic engineering, leveraging the principles of artificial magnetic conductors (AMCs) discussed previously. PRGW technology has received considerable attention in recent years, primarily due to its demonstrated low losses in the millimeter/terahertz-wave spectrum [21–24]. Unlike traditional waveguides, the PRGW structure shown in Figure 1 is constructed on printed circuit boards (PCBs), with the wave propagating over an air gap where dielectric losses are minimal. This remarkable design aspect significantly mitigates transmission loss, ensuring efficient signal propagation with minimal distortion and preserving signal integrity, crucial for high-frequency applications. Leveraging the quasi-TEM mode, which exhibits lower dispersion, minimizes degradation of signal quality, particularly important for maintaining signal transmission without distortion [21,22]. Its utilization in fields such as passive devices and antenna systems exemplifies the significance of microwave engineering in modern telecommunications [20,25–27].

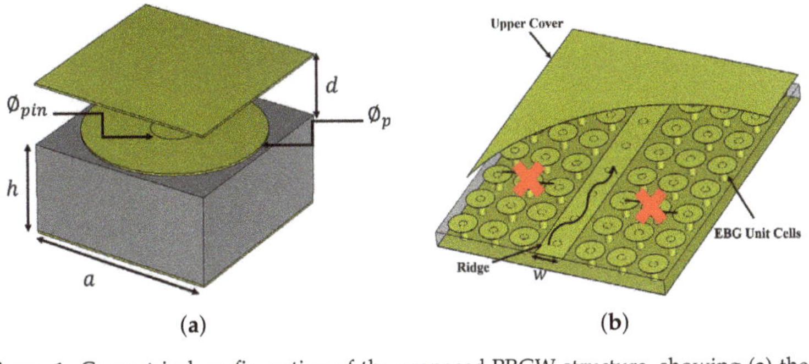

Figure 1. Geometrical configuration of the proposed PRGW structure, showing (**a**) the unit cell and (**b**) the PRGW, with dimensions a = 1.6 mm, h = 0.762 mm, ϕ_{pin} = 1.5 mm, d = 0.5 mm and w = 1.37 mm.

The construction of the Printed Ridge Gap Waveguide (PRGW) involves a periodic arrangement of unit cells on a printed circuit board (PCB). Each unit cell, as shown in Figure 1a, consists of an artificial magnetic conductor (AMC) surface covered by a ground plane on the top. This configuration creates a band-gap where wave propagation is inhibited within a specific frequency range. However, by introducing a ridge between these periodic electromagnetic band-gap (EBG) cells, wave propagation is supported within the band-gap, enabling efficient signal transmission. The cutoff frequency is typically defined for the PRGW line rather than for the individual unit cells because the insertion of the ridge reduces the band-gap compared to the bandwidth of the band-gap of the periodic unit cells. Inserting the ridge disrupts the infinite periodicity assumed when calculating the band-gap bandwidth. Traditionally, the cutoff frequency and operating bandwidth of the PRGW are determined using eigenmode solvers in simulation tools. By appropriately defining the boundary conditions, the dispersion diagram can be extracted, and the bandwidth can be obtained along with the cutoff frequency value [28]. This conventional approach requires extensive use of simulation tools to calculate the initial dimensions needed to achieve the desired cutoff frequency and bandwidth. This process is often time-consuming and relies on a trial-and-error method, which lacks physical intuition and efficiency. In contrast, this work proposes an analytical procedure to calculate the cutoff frequency of PRGW. This method is based on the equivalent circuit model of the mushroom cell and its design procedure [29]. By employing this analytical approach, we aim to provide a more efficient and insightful method for determining the cutoff frequency and bandwidth, reducing dependency on simulation tools and expediting the design process.

Building on the previous discussion of the PRGW structure and its performance, the focus now shifts to the equivalent circuit model of the Electromagnetic Band-Gap (EBG) structure, particularly the mushroom-shaped unit cell depicted in Figure 1. Our analytical approach, which is essential for accurately predicting the behavior of PRGW systems, relies on this circuit model. The primary objective here is to determine the cutoff frequency of the PRGW analytically and to validate these results through comparison with simulated data. Traditionally, this determination is accomplished by plotting the dispersion diagram using CST and extracting the cutoff frequency via the eigenmode solver, as outlined earlier. To understand this process, consider that when an electromagnetic wave impinges on an array of mushroom-shaped unit cells, it induces electric fields across the gaps between the cells. These fields can be effectively modeled as an equivalent capacitance, denoted as C. Additionally, the incident fields generate currents that flow between adjacent unit cells, creating current paths through the walls or vias. This behavior can be represented by an equivalent inductance, L. Consequently, the equivalent circuit model of each unit cell comprises a parallel combination of capacitance C and inductance L, as shown in Figure 2 [29,30].

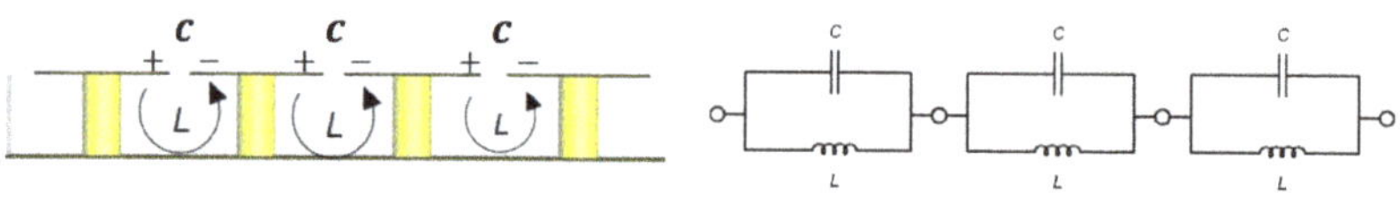

(a) (b)

Figure 2. Geometrical configuration of the AMC surface and its equivalent circuit model. (**a**) Unit cell representing the structural elements, and (**b**) an equivalent circuit model showing the electrical counterparts.

The surface impedance of such a structure can be calculated using the following equation:

$$Z_s = \frac{j\omega L}{1 - \omega^2 LC} \tag{1}$$

where the resonance frequency is:

$$\omega_o = \frac{1}{\sqrt{LC}} \tag{2}$$

The previously mentioned inductance and capacitance account for both the geometry of the individual unit cells and their overall geometric configuration. Using these parameters, the capacitance of each unit cell can be calculated as follows [29]:

$$C = \frac{\phi_p \epsilon_o (\epsilon_r + 1)}{\pi\sqrt{3}} \cosh\left(\frac{a}{g}\right)^{-1} \quad (3)$$

where ϵ_o and ϵ_r are the permittivities of the air and the substrate material, respectively, while ϕ_p is the diameter of the circular patch unit cell.

From this capacitance, the inductance can be found using the following relation:

$$L = \frac{1}{C\omega_0^2} \quad (4)$$

where ω_o is the angular resonance frequency, calculated as $\omega_o = 2\pi f_o$, with f_o being the center operating frequency, specified as 30 GHz in our work. With the values for the circuit elements determined, the surface impedance Z_s can be calculated using Equation (1). The reflection phase of the proposed structure can be determined using the following equation [30]:

$$\angle S11 = \Im m\left(\ln\left(\frac{\eta_o(\omega^2 LC - 1) + j\omega L}{\eta_o(1 - \omega^2 LC) + j\omega L}\right)\right) \quad (5)$$

where η_o is the characteristic impedance of air. From the reflection phase, the operating band of the proposed unit cell can be predicted, with the fractional bandwidth lying between $+90°$ and $-90°$. Thus, the cutoff frequency is the lower frequency f_l, corresponding to a phase of $90°$. This analytical procedure is verified using CST-Microwave Studio, as shown in Figure 3a, where the unit cell is simulated as a periodic structure on two main axes. The reflection phase from this simulation is compared with the analytical results, as shown in Figure 3b, demonstrating good agreement between both methods.

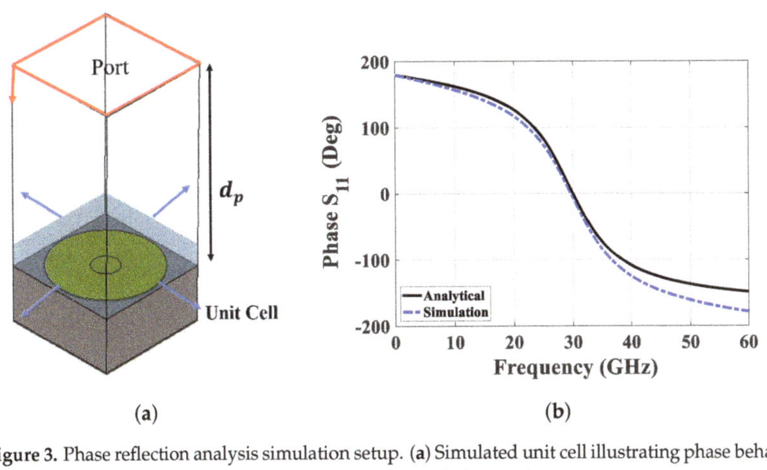

Figure 3. Phase reflection analysis simulation setup. (**a**) Simulated unit cell illustrating phase behavior, and (**b**) a comparison between analytical and simulated phase values.

The full characterization of the PRGW structure requires an examination of its wave propagation phase constant. Wave propagation within the PRGW is enabled by inserting a ridge between the unit cells, extending beyond the EBG structure, as depicted in Figure 1b. This structure allows the wave to propagate in the gap over the ridge, where the propagation constant β serves as the primary parameter governing the phase behavior of the structure. The propagation constant inside the PRGW can be calculated as follows:

$$\beta = \sqrt{k^2 - k_c^2} \quad (6)$$

where $k = \dfrac{2\pi\epsilon_{eff}}{\lambda}$ and $k_c = \dfrac{2\pi}{\lambda_c}$ are the wavenumber and the cutoff wavenumber of the structure, while ϵ_{eff} and λ_c are the effective permittivity and the cutoff wavelength of the PRGW, respectively. This analytical approach allows for determining propagation constants without reliance on CST software or other simulation tools. Understanding these propagation constants is crucial for comprehending the behavior of electromagnetic waves within the PRGW structure. Such insights are invaluable in the design of microwave devices, particularly components like the phase shifter that will be discussed later, where precise control over phase behavior is critical for optimal performance.

It is worth mentioning that the effective permittivity is used instead of the dielectric constant of the PRGW substrate to account for the complex field distribution and interactions within the waveguide structure. The accurate calculation of the effective permittivity ϵ_{eff} is crucial for determining the wavenumber k and, subsequently, the propagation constant β. In previous studies, the effective permittivity has been determined for Inverted Microstrip Gap Waveguide (IMGW) structures [31]. These studies provide an empirical equation for ϵ_{eff} based on the specific geometry and material properties of IMGW. However, this empirical equation cannot be directly applied to PRGW due to distinct structural differences between the two waveguides. These differences include the presence of a ridge in PRGW, while IMGW features a microstrip line on a dielectric substrate.

To bridge the gap between the theoretical and practical calculations of effective permittivity, an adjustment to the empirical equation used for the Inverted Microstrip Gap Waveguide (IMGW) was required. This adjustment process involved deriving a correction factor through a comprehensive non-linear curve fitting procedure. Given that PRGW experiences a reduced dielectric influence compared to IMGW, the effective permittivity (ϵ_{eff}) in PRGW is naturally lower. This crucial insight guided the modification of the empirical equation, ensuring it accurately reflects the effective permittivity for PRGW. The refinement process entailed extensive simulations and analytical comparisons, iteratively adjusting the empirical formula to achieve a close match with the observed data for PRGW. Through this methodical approach, a precise calculation method for ϵ_{eff} in PRGW was developed, which contributes significantly to the literature by providing a tailored solution for this specific waveguide structure. This improved methodology enhances the accuracy of phase constant calculations, essential for the design of microwave components such as phase shifters. The modified equation for the PRGW effective permittivity differs from that of IMGW, exhibiting an exponential decay behavior, and is expressed as follows:

$$\begin{aligned}\epsilon_{PRGW_{eff}} = 1.137e^{-0.07068\epsilon_r} \Bigg[&\frac{\epsilon_r+1}{2}\left(1-0.0004\ (\frac{2W}{d})^2\right) \\ &-\frac{\epsilon_r-1}{2}\left(1+0.2794(\frac{d}{2W})+4.4735(\frac{h}{W})^{-0.3105}\right) \\ &-0.0009\left(1+1.3382(\frac{W}{h})^{1.2748}\right)\Bigg]\end{aligned} \quad (7)$$

Following the refinement of the analytical approach and the derivation of ϵ_{eff} in PRGW, researchers can now accurately determine the propagation constant for the Printed Ridge Gap Waveguide (PRGW). Employing the analytical equation developed earlier, a comprehensive analysis is conducted for the PRGW structure, where the analytical predictions are compared to simulated propagation constants, as illustrated in Figure 4. This comparison reveals a high degree of agreement between the analytical and simulated results, affirming the accuracy of the analytical approach in predicting the propagation behavior of the PRGW structure.

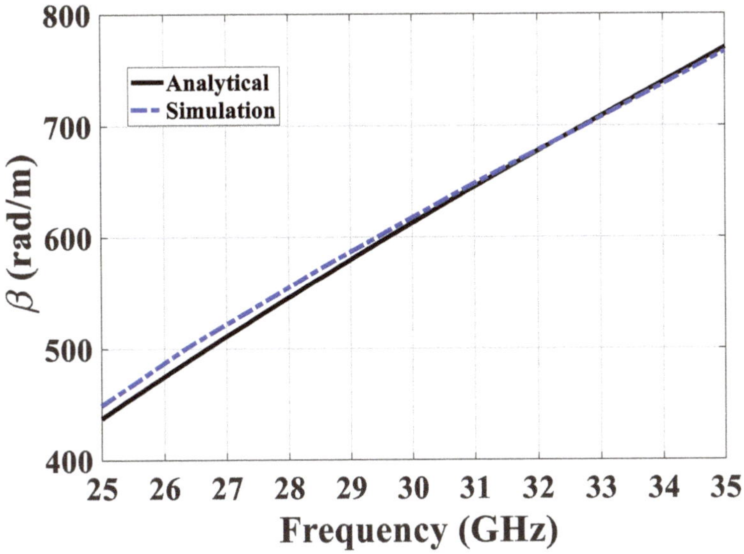

Figure 4. Comparison of analytical and simulated results for PRGW propagation constants, demonstrating the precision of the developed analytical model.

3. Proposed PRGW Phase Shifter: Theoretical Analysis and Design Methodology

In this section, the application of Printed Ridge Gap Waveguide (PRGW) technology is explored for the development of a novel phase shifter. The design of the proposed phase shifter is founded on a directional multi-layer coupler architecture, utilizing the unique characteristics of PRGW. A comprehensive analysis of the coupling mechanism and the differential phase shift within the PRGW structure is undertaken. Following this analysis, the detailed geometry of the proposed design is presented, aimed at overcoming implementation challenges, such as achieving optimal matching and ensuring a flat phase shift response. In conclusion, we provide a concise summary of the design process employed for developing the phase shifter, highlighting key insights and considerations and presenting the simulated results to validate the effectiveness of the proposed design methodology.

3.1. Theoretical Analysis of Multi-Layer PRGW Phase Shifter

In the theoretical analysis of the proposed phase shifter, the fundamental design is described as a two-port device, derived from the modification of a multi-layer patch coupler shown in Figure 5 [32–34]. In this design, two ports are terminated with open circuits, while the remaining two serve as input and output ports. Critical performance indicators, such as the insertion loss and the return loss, are commonly used to evaluate microwave components. However, for a phase shifter, additional considerations are crucial. Among these are the differential phase shift's accuracy and the phase response's flatness. These specifications are essential to ensure that the phase shifter meets the rigorous standards of modern communication systems, particularly in scenarios requiring precise phase control.

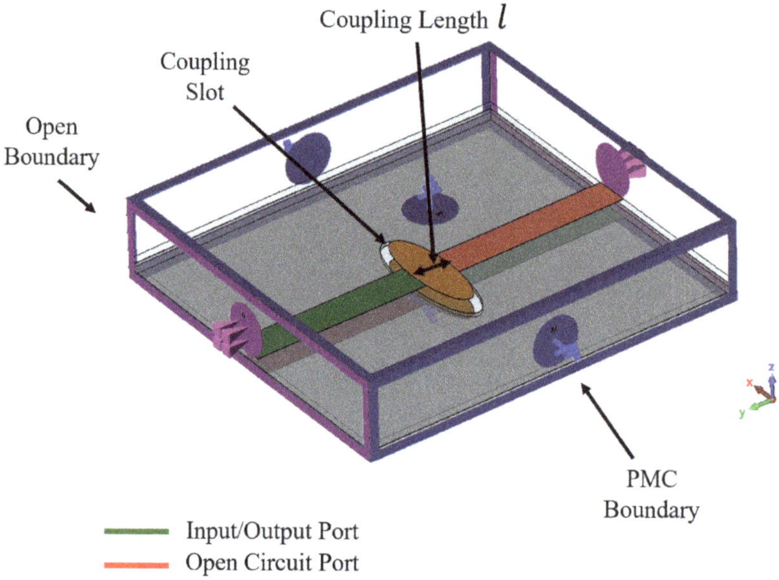

Figure 5. Ideal case for a multi-layer coupler with perfect magnetic conductor and open boundary conditions.

To quantify the phase shift introduced by the proposed network, S-parameter network analysis is employed, assuming a coupling coefficient denoted as C_p. This phase shift is calculated relative to a PRGW reference line, which facilitates the assessment of phase changes caused by the proposed design. The phase shift obtained from the network is determined as follows: [32]:

$$\phi_c = \frac{\pi}{2} - 2\tan^{-1}\left(\frac{\sin(\beta l)}{\sqrt{1 - {C_p}^2}\cos(\beta l)}\right) \tag{8}$$

In the previous equation, the used propagation constant β is obtained from the proposed Equation (6) and l is the physical length of the coupling structure, which is the secondary diameter of the elliptical shape, as shown in Figure 5. The phase shift obtained from the reference PRGW line is calculated as follows:

$$\phi_m = -\beta l_m \tag{9}$$

Therefore, the differential phase shift can be calculated as follows:

$$\Delta\phi_c = \frac{\pi}{2} - 2\tan^{-1}\left(\frac{\sin \beta l}{\sqrt{1 - C_p^2}\cos \beta l}\right) + \beta l_m \tag{10}$$

Following the theoretical analysis, the achieved phase shift for various values of the normalized coupling length βl and the coupling coefficient C_p are depicted in Figure 6. To achieve a wide range of phase shifts from 0° to 135°, the physical length of the reference line l_m is optimized to minimize the deviation in the differential phase shift. The estimated phase range extends from 0° to 135° for C values ranging from 0.45 down to 0.05. However, it is essential to evaluate the phase shifter's operation by examining the return loss and insertion loss. As indicated by the equations in Table 1, the coupling value significantly affects the return loss S_{11} and insertion loss S_{21}. This poses a limitation to our design in achieving high levels of phase shift. Specifically, for large phase shift values, the coupling value increases, leading to degradation in the return loss and, consequently, the insertion loss. Therefore, for large phase shift values, design modifications and tuning elements for matching are necessary, as will be discussed in the following section.

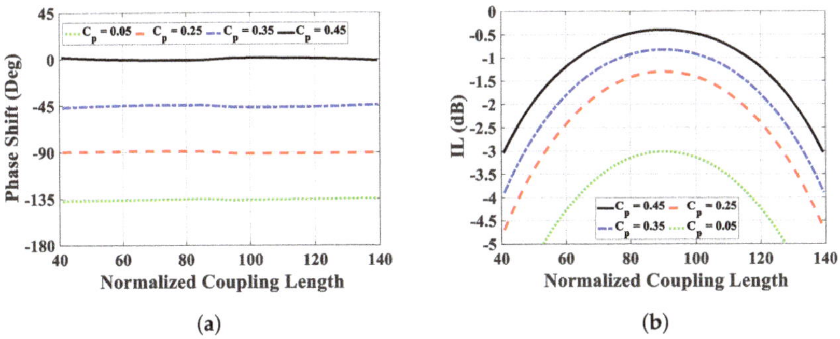

Figure 6. Phase shift and insertion loss variation due to changes in coupling parameters. (**a**) Phase shift, and (**b**) insertion loss.

Table 1. S-parameters equations depending on the even-odd analysis of multi-layer coupler [32].

Quantity	Expression
S_{11}	$\dfrac{1 - C_p(1 + \sin^2(\beta l))}{\left[\sqrt{1 - C_p{}^2}\cos(\beta l) + j\sin^2(\beta l)\right]^2}$
S_{21}	$\dfrac{2jC\sqrt{1 - C_p{}^2}\sin(\beta l)}{\left[\sqrt{1 - C_p{}^2}\cos(\beta l) + j\sin^2(\beta l)\right]^2}$

3.2. *Proposed Multi-Layer PRGW Phase Shifter Design*

The proposed configuration of the multi-layer PRGW phase shifter is illustrated in Figure 7a. This design primarily relies on a coupling section consisting of two elliptical patches positioned on the top and bottom layers. These patches are interconnected with the input and output PRGW lines. Proximity coupling between the patches is facilitated by cutting an elliptical slot in the ground planes of a thin RT6002 substrate with a thickness of H_2. Plated vias are placed around the coupling aperture to confine the electromagnetic field within the slot. The geometric parameters critical to the performance of the coupler include the width W_p and length L_p of the elliptical patches, as well as the width W_s and length L_s of the elliptical coupling slot. Additionally, a cut in the slot is included with dimensions w_c and l_c to provide another degree of freedom for fine control of the phase shift without affecting the matching level. These additional parameters are crucial for optimizing the phase shift independently of the matching level, ensuring robust performance of the phase shifter. All of these parameters are meticulously designed to ensure optimal coupling efficiency and minimal signal loss, thereby enhancing the overall performance of the phase shifter.

To address the theoretical analysis discussed previously, the design of the elliptical patches and slots ensures the desired coupling and differential phase shift. However, as indicated in the theoretical analysis, higher values of phase shift can degrade the return loss and insertion loss. To mitigate this, a matching transformer is incorporated into the proposed phase shifter, as shown in Figure 7b. This matching transformer consists of two lines with different widths to match the elliptical shape impedance to the 50 Ω line. In addition to the matching transformers, another modification to the multi-layer phase shifter is proposed to improve the matching level and add a degree of freedom to achieve the required phase shifter values while maintaining a deep matching level. A rectangular patch with dimensions l_r and w_r is added, and the elliptical slot is modified with a rectangular cut with dimensions l_c and w_c, as shown in Figure 7b. These modifications help achieve phase shifters up to 135°, maintaining a deep matching level over the entire bandwidth, as will be demonstrated in the following section.

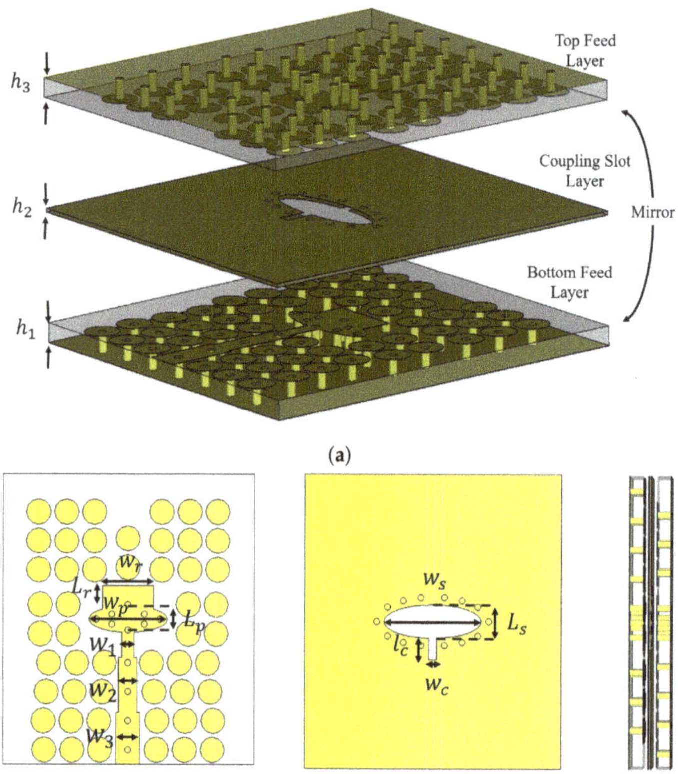

Figure 7. The geometry of the proposed PRGW phase shifter design. (**a**) 3D view, and (**b**) cross-sectional view.

3.3. Design Methodology and Simulation Results

The phase shifter is developed using a multi-layer coupler design, with two ports terminated as open circuits and the other two serving as input and output connections. The coupling mechanism from the bottom layer to the top layer follows the approach used in the multi-layer coupler proposed in [35]. The even and odd mode analysis conducted in that study is utilized to determine the initial dimensions for the phase shifter. The design methodology begins by selecting the required coupling value C_p from the previously generated curves. These curves guide the design process for determining the values of W_p and W_s [35]. These values are subsequently optimized for each type of phase shifter, and the final tuned values are provided in Table 2. Moreover, the length of the elliptical patch L_p is chosen to be equal to a quarter of the effective wavelength $\lambda_{eff} = \lambda/\sqrt{\epsilon_{eff}}$ at the center frequency of operation. Here, in our case, it is found to be around 1.5 mm, as depicted in Table 2. Once the necessary dimensions are established, the phase shift is compared to the reference line, as illustrated in Figure 8. This reference line is designed in a curved manner to ensure a flat phase response. The dimensions of the coupling slot and elliptical patch are fine-tuned along with other coupler parameters to achieve the desired performance. To achieve a good matching level and low insertion loss, particularly for low coupling values, a matching transformer is integrated into the coupling slot, as shown in Figure 7. Additionally, the length of the line l_{line} is optimized to achieve the various phase shifts with a flat response. Simulation results are presented in Figure 9, where it is evident that the matching level for the proposed design is below 20 dB for all phase shift cases. Furthermore, the flatness of the phase shift curves is maintained within $\pm 5^\circ$ of the designed value (0° to 135°), demonstrating the effectiveness of the proposed design methodology. These simulation results underscore the capability of the proposed phase shifter to deliver accurate phase shifts with excellent matching and low insertion loss, fulfilling the stringent requirements of modern communication systems.

Table 2. Phase shifter dimensions for different values of phase shift.

Parameter	0° Phase	90° Phase	135° Phase
W_p (mm)	4.57	3.67	3.75
ine L_p (mm)	1.48	1.55	1.5
W_s (mm)	5.53	6.9	7
ine L_s (mm)	1.8	1.53	1.1
L_{line} (mm)	4.9	2.9	2

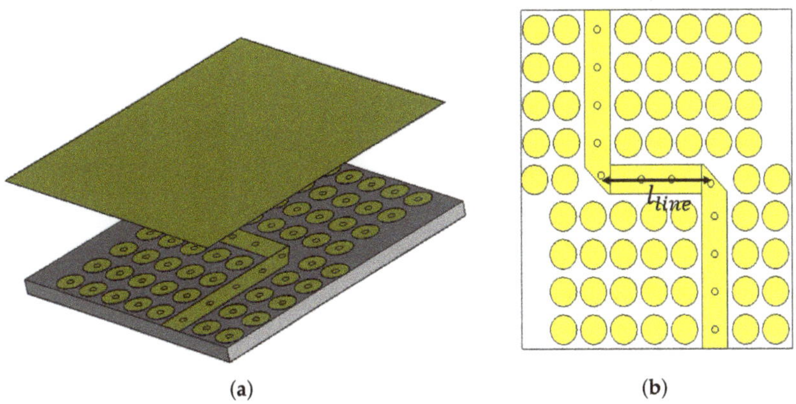

Figure 8. The geometry of the proposed PRGW reference line for phase comparison. (**a**) 3D view, and (**b**) cross-sectional view.

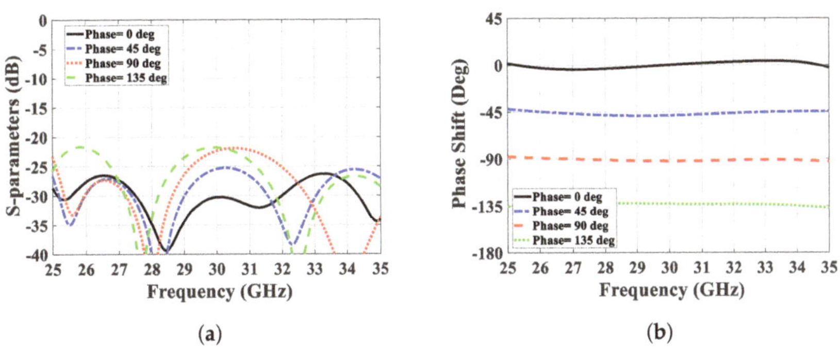

Figure 9. Phase shifter simulation results for different phase values. (**a**) S-parameter, and (**b**) differential phase.

4. Measurements and Validation

The proposed PRGW phase shifter undergoes experimental validation through fabrication and measurement, as depicted in Figure 10a. We specifically select a 45° phase shifter for validation, with detailed dimensions provided in Table 3. Fabrication involves assembling the PRGW phase shifter components using high-temperature and high-pressure epoxy. To ensure precise performance evaluation, we utilize a TRL calibration kit to mitigate any effects arising from connectors and microstrip line transitions. The S-parameters are measured using the ANRITSU MS46322A VNA, and a comparative analysis between the measured and simulated results is presented in Figure 10b. Notably, there is strong agreement between the measured and simulated S-parameters, achieving a relative bandwidth of 38% at 30 GHz. However, some discrepancies are observed in the measured results at the first of the band, primarily attributable to fabrication tolerances and the adhesive used during assembly. Furthermore, we evaluate the phase shift obtained from the phase shifter, comparing it with the reference line and simulated results, as depicted in Figure 10c. The proposed phase shifter demonstrates a phase balance of approximately $45 \pm 5°$ across the entire operational bandwidth. The proposed phase shifter not only demonstrates impressive phase balance but also exhibits desirable matching characteristics. The measured insertion loss remains below -0.5 dB, ensuring minimal signal attenuation. Additionally, the matching level achieved is below -16 dB, indicating efficient power transfer and minimal signal reflection within the operating bandwidth.

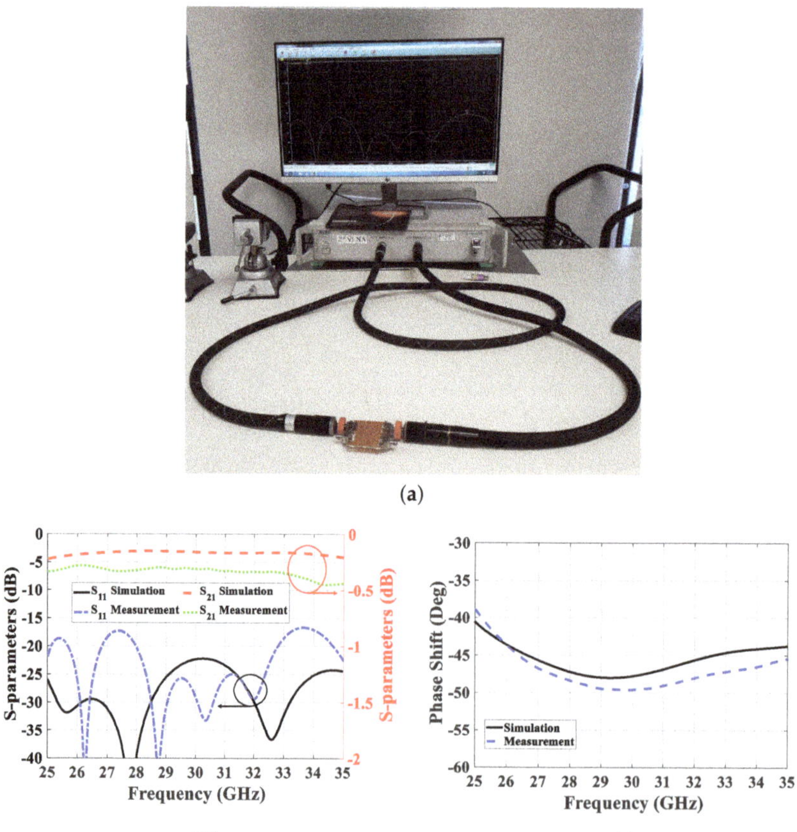

Figure 10. Phase shifter measurements verification with the simulation results. (**a**) Measurement setup, (**b**) S-parameter, and (**c**) differential phase.

Table 3. 45° phase shifter dimensions.

Parameter	Value (mm)	Parameter	Value (mm)
h_1	0.762	l_{line}	3.6
h_2	0.127	W_1	0.7
W_2	1.05	W_3	1.377
W_p	4.5	L_P	1.3
W_r	0.762	L_r	1.3
W_s	5.5	L_s	1.85
W_c	0.48	L_c	1.17

As we discuss the measurement and validation of our proposed multi-layer PRGW phase shifter, it is important to compare our work with the existing literature. While there is considerable research on phase shifters, particularly in millimeter-wave applications, the utilization of PRGW remains relatively limited. This highlights the unique contribution of our work, which seeks to address this under-explored area of research. By comparing our results with state-of-the-art methodologies documented in prior studies, we aim to underscore the advancements and distinctiveness of our PRGW-based phase shifter. The comparison Table 4 offers a thorough examination of the performance characteristics across various configurations of Ridge Gap Waveguide (RGW) phase shifters, providing insights into the existing literature in this field. Previous studies have primarily focused on Inverted Microstrip Gap Waveguide (IMGW) and Metallic Ridge Gap Waveguide (GGW) technologies, utilizing techniques such as Schiffman phase shifters [7] and time delay methods [8,9]. Schiffman phase shifters have been effectively utilized to construct linear Butler matrices and implement 1D beam scanning antenna arrays. However, these implementations face challenges due to their narrow operational bandwidth (20%) and large size [7]. These challenges are attributed to the reliance on coupled line theory, which necessitates small gaps between the coupled lines to achieve a wide range of phase shifts. Fabrication limitations make it difficult to achieve these small gaps while maintaining a reasonable matching level, particularly at millimeter-wave frequencies. Similarly, time delay techniques, including those used in PMC packaged or Inverted Microstrip Line Gap Waveguides, have been explored as alternatives to PRGW. However, these methods require additional dielectric layers, resulting in increased costs and complexity, while exhibiting significant phase imbalances over limited bandwidths (around 16%) [8,9]. In contrast, metallic RGW and metallic GGW technologies, while offering promising performance characteristics, face challenges such as high fabrication costs and bulky structures limiting their practical applicability [11,12,36,37]. Conversely, our investigation introduces a new perspective by exploring the application of PRGW in phase shifter designs. This change in focus is notable given the relatively limited exploration of PRGW-based approaches, despite their potential advantages. Unlike phase shifter designs utilizing IMGW and GGW technologies, which often encounter challenges, such as limited bandwidth and less-than-optimal phase balance, our PRGW-based approach offers distinct advantages. While some existing designs achieve better than ±5° degrees phase balance, typically around ±4°, they often come with narrower bandwidth and larger size compared to our proposed design. Our approach using PRGW technology achieves a wider bandwidth and compact size while maintaining competitive phase balance performance. This trade-off is due to the optimized configuration of PRGW, which balances between phase performance, bandwidth, and size constraints. Specifically, our phase shifter designs leverage PRGW's inherent benefits, including enhanced matching levels and phase stability. This focus on PRGW-based designs enables us to overcome many of the limitations associated with IMGW and GGW technologies, ultimately resulting in improved performance metrics. Furthermore, the proposed work emphasizes the significance of multi-layer configurations as a promising approach for compact beamforming networks. Notably, the prior literature has largely overlooked this aspect. Only a single previous study has investigated multi-layer phase shifters based on IMGW technology, with limited bandwidth and performance metrics [9]. In contrast, our innovative design methodology enables the realization of a broad range of phase shifts from 0° to 135°, while ensuring a deep matching level across the entire spectrum. This capability represents a significant advancement over existing methodologies, which often struggle to achieve comparable levels of performance.

Table 4. Performance comparison between different RGW phase shifter configurations.

Ref.	RL [dB]	IL [dB]	f_o (GHz)	BW %	Ph. Balance	Size (λ_o^2)	Tech.
[7]	14	1	30.71	20	$45 \pm 3.5°$	-	PRGW
[9]	-	-	30	16.7	$45 \pm 3°$	1×2	IMGW
[8]	-	0.6	29	13.8	$45 \pm 3°$	0.6×0.3	IMGW
[12]	10	0.7	26	26.3	$85 \pm 5°$	2.5×3.6	RGW
[11]	15	-	27	29.63	$90 \pm 4°$	2.9×4.6	RGW
[36]	25	-	94	6.4	$90 \pm 4°$	2.6×3.2	GGW
[37]	12	0.75	69.5	15.8	-	-	GGW
This Work	16	0.5	30	33.33	$45 \pm 5°$	1×1	PRGW

5. Conclusions

In this paper, we have introduced compact Printed Ridge Gap Waveguide (PRGW) phase shifters designed for millimeter-wave applications, focusing on achieving wide operating bandwidth, improved matching, and enhanced phase balance compared to single-layer technology. A unique methodology has been presented for achieving the required phase shift in PRGW technology, which has not been previously explored. Additionally, a novel analytical approach to calculate the cutoff frequency and propagation constant of the PRGW structure has been introduced, providing a new method to characterize PRGW structures. By employing multi-layer PRGW technology, we have enabled the realization of multi-layer beamforming networks without crossing, supporting wideband operation in a compact form factor. Our design procedure has successfully achieved phase shift values ranging from 0° to 135° over a broad frequency bandwidth centered at 30 GHz. A 45-degree phase shifter has been fabricated and measured, demonstrating a 10 GHz bandwidth (approximately 33% fractional bandwidth) from 25 GHz to 35 GHz, with a phase balance maintained within $45 \pm 5°$ and a deep matching level of -20 dB. These results underscore the superior performance of the proposed phase shifter in terms of bandwidth, phase balance, and matching levels, making it a promising candidate for advanced millimeter-wave applications, including beyond 5G (B5G) and 6G wireless communications. The proposed design, capable of implementing a wide range of phase shifts, is particularly advantageous for developing scalable beamforming networks such as 4×4 and 8×8 Butler matrices, accommodating the varying phase shift values required for larger networks.

Author Contributions: Conceptualization, M.S.E.-G. and M.M.M.A.; data curation, M.A. and M.G.; funding acquisition, M.A.; investigation, M.M.M.A. and M.G.; methodology, M.M.M.A.; formal analysis, M.S.E.-G.; resources, M.A.; software, M.S.E.-G. and M.M.M.A.; validation, M.G.; writing—original draft, M.G.; writing—review and editing, M.S.E.-G. All authors have read and agreed to the published version of the manuscript.

Funding: This work was supported by King Saud University, Riyadh, Saudi Arabia, through the Researchers Supporting Project under Grant RSPD2024R868.

Data Availability Statement: Data are contained within the article.

Acknowledgments: The authors would like to thank King Saud University, Riyadh, Saudi Arabia for the support provided. Also, the authors would like to thank the Electronics Research Institute (ERI), Cairo, Egypt for the support provided.

Conflicts of Interest: Author Mahmoud Gadelrab was employed by the company Scientific Microwave Corporation. The remaining authors declare that the research was conducted in the absence of any commercial or financial relationships that could be construed as a potential conflict of interest.

References

1. Wang, C.X.; You, X.; Gao, X.; Zhu, X.; Li, Z.; Zhang, C.; Wang, H.; Huang, Y.; Chen, Y.; Haas, H.; et al. On the Road to 6G: Visions, Requirements, Key Technologies, and Testbeds. *IEEE Commun. Surv. Tutorials* **2023**, *25*, 905–974. [CrossRef]
2. Semaan, E.; Tejedor, E.; Kumar-Kochhar, R.; Magnusson, S.; Parkvall, S. 6G Spectrum—Enabling the Future Mobile Life beyond 2030. White Paper GFTL-23:000243 Uen. 2023. Available online: https://www.ericsson.com/en/reports-and-papers/white-papers/6g-spectrum-enabling-the-future-mobile-life-beyond-2030 (accessed on 18 June 2024).
3. Alwahishi, R.; Ali, M.M.M.; Elzwawi, G.; Denidni, T.A. Beam-switching antenna using reconfigurable intelligent frequency selective surfaces for internet of things applications. *IEEE Internet Things J.* **2023**, *11*, 4152–4162. [CrossRef]
4. ITU-R. Recommendation ITU-R M.2160-0: Framework and Overall Objectives of the Future Development of IMT for 2030 and Beyond. M Series: Mobile, Radiodetermination, Amateur and Related Satellite Services. 2023. Available online: https://www.itu.int/rec/R-REC-M.2160-0-202311-I/en (accessed on 18 June 2024).
5. Tataria, H.; Shafi, M.; Molisch, A.F.; Dohler, M.; Sjöland, H.; Tufvesson, F. 6G Wireless Systems: Vision, Requirements, Challenges, Insights, and Opportunities. *Proc. IEEE* **2021**, *109*, 1166–1199. [CrossRef]
6. Rabbanifard, M.; Zarifi, D.; Farahbakhsh, A.; Mrozowski, M. Design of Compact and Wideband Groove Gap Waveguide-Based Directional Couplers. *IEEE Access* **2024**, *12*, 86346–86354. [CrossRef]
7. Afifi, I.; Sebak, A.R. Wideband 4 × 4 Butler Matrix in the Printed Ridge Gap Waveguide Technology for Millimeter-Wave Applications. *IEEE Trans. Antennas Propag.* **2020**, *68*, 7670–7675. [CrossRef]
8. Shady, M.O.; Kishk, A.A. Compact 4 × 4 Multilayer Butler Matrix With Four-Slot Array. *IEEE Trans. Microw. Theory Tech.* **2023**, *72*, 3522–3529. [CrossRef]
9. Ashraf, N.; Sebak, A.R.; Kishk, A.A. PMC Packaged Single-Substrate 4 × 4 Butler Matrix and Double-Ridge Gap Waveguide Horn Antenna Array for Multibeam Applications. *IEEE Trans. Microw. Theory Tech.* **2021**, *69*, 248–261. [CrossRef]
10. Abdelaal, M.A.; Shams, S.I.; Kishk, A.A. Compact RGW Differential Phase Shifter for Millimeter-Wave Applications. In Proceedings of the 2018 18th International Symposium on Antenna Technology and Applied Electromagnetics (ANTEM), Waterloo, ON, Canada, 19–22 August 2018; pp. 1–2. [CrossRef]
11. Ali, M.M.M.; Patel, S.; Sebak, A.R. A Novel Wideband Ridge Gap Waveguide Magic Tee for Millimeter-Wave Application. In Proceedings of the 2022 International Telecommunications Conference (ITC-Egypt), Alexandria, Egypt, 26–28 July 2022; pp. 1–6. [CrossRef]
12. Palomares-Caballero, Á.; Megías, C.; Molero, C.; Alex-Amor, A.; Padilla, P. Wideband Gap-Waveguide Phase Shifter Based on a Glide-Symmetric Ridge. *IEEE Microw. Wirel. Technol. Lett.* **2023**, *33*, 27–30. [CrossRef]
13. Luukkonen, O. Artificial Impedance Surfaces. 2009. Available online: https://core.ac.uk/download/pdf/301128762.pdf (accessed on 18 June 2024).
14. Rahmat-Samii, Y.; Mosallaei, H. Electromagnetic Band-Gap Structures: Classification, Characterization, and Applications. In Proceedings of the 2001 Eleventh International Conference on Antennas and Propagation, Manchester, UK, 17–20 April 2001.
15. Mosallaei, H.; Rahmat-Samii, Y. Periodic bandgap and effective dielectric materials in electromagnetics: Characterization and applications in nanocavities and waveguides. *IEEE Trans. Antennas Propag.* **2003**, *51*, 549–563. [CrossRef]
16. Kildal, P.S.; Zaman, A.U.; Rajo-Iglesias, E.; Alfonso, E.; Valero-Nogueira, A. Design and experimental verification of ridge gap waveguide in bed of nails for parallel-plate mode suppression. *IET Microwaves Antennas Propag.* **2011**, *5*, 262–270. [CrossRef]
17. Kildal, P.S.; Alfonso, E.; Valero-Nogueira, A.; Rajo-Iglesias, E. Local metamaterial-based waveguides in gaps between parallel metal plates. *IEEE Antennas Wirel. Propag. Lett.* **2008**, *8*, 84–87. [CrossRef]
18. Gadelrab, M.; Shams, S.I.; Elsaadany, M.; Sebak, A. Ridge Gap Waveguide Low Pass Filters: A Systematic Design Approach. *IEEE Access* **2024**, *12*, 81992–82001. [CrossRef]
19. Shady, M.O.; Sifat, S.M.; Ali, M.M.M. Exploring Gap Waveguide Solutions: A Review of Millimeter-Wave Beamforming Components and Antennas for Advanced Applications. *J. Electron. Electr. Eng.* **2024**, *3*, 236–254. [CrossRef]
20. Gadelrab, M.; Shams, S.I.; Sebak, A.R. Low Profile Corrugated Horn for Minimum Side lobe Levels based on PRGW. In Proceedings of the 2023 International Microwave and Antenna Symposium (IMAS), Cairo, Egypt, 7–9 February 2023; pp. 243–246.
21. Sifat, S.M.; Ali, M.M.M.; Shams, S.I.; Sebak, A.R. High gain bow-tie slot antenna array loaded with grooves based on printed ridge gap waveguide technology. *IEEE Access* **2019**, *7*, 36177–36185. [CrossRef]
22. Sorkherizi, M.S.; Kishk, A.A. Fully printed gap waveguide with facilitated design properties. *IEEE Microw. Wirel. Components Lett.* **2016**, *26*, 657–659. [CrossRef]
23. Ali, M.M.M.; Shams, S.I.; Elsaadany, M.; Gagnon, G.; Wu, K. Graphene-based terahertz reconfigurable printed ridge gap waveguide structure. *Sci. Rep.* **2022**, *12*, 21111. [CrossRef] [PubMed]
24. Kiani, N.; Tavakkol Hamedani, F.; Rezaei, P. Design of a graphene-based ridge gap waveguide coupler for THz applications. *Opt. Quantum Electron.* **2024**, *56*, 423. [CrossRef]
25. Mousavirazi, Z.; Ali, M.M.M.; Rezaei, P.; Sebak, A.R.; Denidni, T.A. A novel low-loss planar PRGW crossover design for 5G applications. *Radio Sci.* **2023**, *58*, 1–8. [CrossRef]

26. Allam, A.; Mahmoud, A.O.; Fawzy, D.E.; Askar, M. Ultra Wide-Band PRGW Based Magneto-Electric Dipole Antenna. In Proceedings of the 2021 15th European Conference on Antennas and Propagation (EuCAP), Dusseldorf, Germany, 22–26 March 2021; pp. 1–5.
27. Mousavirazi, Z.; Akbari, M.; Denidni, T. Millimeter-wave high-gain PRGW antenna using a Fabry-Perot cavity. In Proceedings of the 2020 IEEE International Symposium on Antennas and Propagation and North American Radio Science Meeting, Montreal, QC, Canada, 5–10 July 2020; pp. 1365–1366.
28. Shams, S.I.; Ali, M.M.M.; Sebak, A.R. Reconfigurable guiding structure based on printed ridge gap waveguide technology. In Proceedings of the 2018 18th International Symposium on Antenna Technology and Applied Electromagnetics (ANTEM), Waterloo, ON, Canada, 19–22 August 2018; pp. 1–2.
29. Balanis, C.A. *Advanced Engineering Electromagnetics*; John Wiley & Sons: Hoboken, NJ, USA, 2012.
30. Sievenpiper, D.F. *High-Impedance Electromagnetic Surfaces*; University of California: Los Angeles, CA, USA, 1999.
31. Hassan, A.T.; Moharram, M.A.; Kishk, A.A. Empirical analysis formulae of microstrip ridge gap waveguide. In Proceedings of the 2018 IEEE International Symposium on Antennas and Propagation & USNC/URSI National Radio Science Meeting, Boston, MA, USA, 8–13 July 2018; pp. 423–424.
32. Abbosh, A.M. Ultra-wideband phase shifters. *IEEE Trans. Microw. Theory Tech.* **2007**, *55*, 1935–1941. [CrossRef]
33. Riblet, H.J. A mathematical theory of directional couplers. *Proc. IRE* **1947**, *35*, 1307–1313. [CrossRef]
34. Oliver, B.M. Directional electromagnetic couplers. *Proc. IRE* **1954**, *42*, 1686–1692. [CrossRef]
35. Ali, M.M.M.; Haraz, O.M.; Afifi, I.; Sebak, A.R.; Denidni, T.A. Ultra-wideband compact millimeter-wave printed ridge gap waveguide directional couplers for 5G applications. *IEEE Access* **2022**, *10*, 90706–90714. [CrossRef]
36. Tamayo-Domínguez, A.; Fernández-González, J.M.; Sierra-Castañer, M. 3-D-printed modified Butler matrix based on gap waveguide at W-band for monopulse radar. *IEEE Trans. Microw. Theory Tech.* **2019**, *68*, 926–938. [CrossRef]
37. Palomares-Caballero, A.; Alex-Amor, A.; Escobedo, P.; Valenzuela-Valdés, J.; Padilla, P. Low-loss reconfigurable phase shifter in gap-waveguide technology for mm-wave applications. *IEEE Trans. Circuits Syst. II Express Briefs* **2020**, *67*, 3058–3062. [CrossRef]

sensors

MDPI

Article

Compact Integrated On-Chip MIMO Antenna with Reconfigurability for mmWave Frequencies

Khaled Boubekeur [1,*], Nicolas Zerounian [1] and Badr Eddine Ratni [2,*]

1 Centre de Nanosciences et de Nanotechnologies (C2N), CNRS, Université Paris-Saclay, F91120 Palaiseau, France; nicolas.zerounian@universite-paris-saclay.fr
2 LEME, UPL, Université Paris Nanterre, F92410 Ville d'Avray, France
* Correspondence: khaled.boubekeur@c2n.upsaclay.fr (K.B.); badreddine.ratni@parisnanterre.fr (B.E.R.)

Abstract: This paper presents a compact on-chip multiple-input multiple-output (MIMO) antenna designed for future communication systems, featuring frequency-agile elements. The antenna achieves enhanced decoupling and reduced cross-section through the integration of a metasurface, which also introduces frequency agility. Designed for the millimeter-wave band using low-loss BenzoCycloButene (BCB) polymer, the antenna is manufactured with microelectronic processes, and the dimensions are $7.54 \times 7.54 \times 0.055$ mm^3. Simulations and measurements demonstrate excellent frequency agility around 60 GHz, with gains of 6.5 to 9 dBi. As a proof of concept, open and short circuits were used for switching, with future designs aiming to incorporate diodes for a full dynamic reconfiguration. This work highlights the potential for compact, high-performance, and frequency-reconfigurable on-chip antennas in next-generation millimeter-wave systems.

Keywords: on-chip antennas; MIMO; artificial magnetic conductor; BenzoCycloButene; metasurface; microelectronic processes

1. Introduction

The rapid advancement of digital technology has driven a dramatic increase in data generation [1], requiring ever-higher data rates to enable efficient communication. Current wireless systems operating in frequency bands from MHz to tens of GHz have limited transmission capacity and may not meet future requirements. To address this challenge, emerging communication systems targeted higher frequency bands, such as 5G (around 30 GHz) and WiGig (Wi-Fi around 60 GHz), which offer greater transmission capacity [2–5]. The FCC has opened up the terahertz spectrum for potential 6G applications, promising significant data rates due to the wide channel bandwidths available at these frequencies [6].

However, the shift to higher frequencies brings challenges, particularly shorter wavelengths, which necessitate the miniaturization and integration of RF systems to minimize interconnection losses and mitigate the high atmospheric attenuation at these frequencies [7]. Today, the literature features numerous studies on highly integrated RF systems and circuits on chips, particularly in the mmWave and THz ranges, such as complete transceiver systems [8]. In this context, integrated antennas are indispensable for achieving compact and efficient wireless communication systems. Their design must simultaneously address multiple performance requirements, including wide bandwidth, reconfigurability, and advanced functionalities like diversity (e.g., MIMO systems) while being seamlessly integrated into transceiver chips [9–11].

In this article, we propose the design of an integrated antenna system for 60 GHz Wi-Fi. The system features MIMO antennas with frequency reconfiguration capabilities

https://doi.org/10.3390/s25041062

to effectively cover a wide frequency band. To meet these demanding requirements, metasurface-based antennas present a promising solution. Metasurfaces, derived from metamaterials, are engineered by arranging small scatterers or openings in a precise pattern on a 2D surface. Over the past decade, they have evolved from theoretical constructs to commercially viable technologies [12,13]. These structures have revolutionized antenna design by overcoming the limitations of traditional antennas, enabling enhanced directivity, reconfigurability, and low losses. They can therefore be leveraged to develop even more efficient integrated antenna systems [14].

In this work, we present an on-chip four-port MIMO antenna that integrates the benefits of metasurface-based isolation, compactness, and frequency agility, integrated onto BenzoCycloButene (BCB) as a substrate [15]. Available in resin for microelectronic fabrication processes, this polymer offers advantageous characteristics, including low losses (tg δ = 0.0008) and low dielectric constants (εr = 2.65) at frequencies in the tens of GHz, as well as at 1 THz (tg δ = 0.008 and ε_r = 2.45) [16]. These properties contribute to improving antenna efficiency and facilitate the extension of the design to the submillimeter-wave range, as already achieved for unimodal coplanar waveguides up to 760 GHz [17]. Herein, our design targets Wi-Fi (802.11ay) applications at 60 GHz [18]. By combining metasurface technology with BCB integration, our approach delivers an optimal balance between size, performance, and reconfigurability, positioning it as a strong candidate for the next generation of wireless communication systems.

2. Antenna Design and Simulations

The idea is to design a compact MIMO antenna consisting of 4×4 elements which can be directly integrated onto an electronic chip for future communication systems. A frequency band around 60 GHz has been selected as the operating band for the development of this antenna, intended for future communication systems requiring, for example, two separate bands for uplinking and downlinking. The antenna is designed using microelectronic fabrication processes, with BenzoCycloButene (BCB) resin to cure the dielectric substrate and gold layer deposition for each metal level. BCB is chosen to minimize dielectric losses even at 60 GHz (εr = 2.65 and tg δ = 0.0008). BCB requires only 250 °C for a complete crosslinking compared to polyimide of a little higher permittivity with almost 350 °C. With a thickness ranging from few microns to few tens of microns, such a dielectric allows for monolithic IC designs of small dimensions matched to the sub-mm wavelength range compared to PCBs. For this purpose, the proposed MIMO antenna structure is based on a conventional linearly polarized patch antenna positioned above a metasurface, as shown in Figure 1. The metasurface serves two key functions. First, it enhances compactness by acting as an artificial magnetic conductor (AMC). This allows for a significant reduction in the total thickness of the substrate between the antenna and the ground plane to only 55 μm with the metasurface compared to the $\lambda 0/4$ = 770 μm without a metasurface for the chosen polymer at 60 GHz with ε_r = 2.65. The metasurface achieves this by forcing the phase of the reflected E field to 0° at the patch level, as shown in Figure 2c,d. Secondly, the metasurface is within the frequency agility by the integration of active components in it. This frequency agility, in addition to broadening the operational bandwidth and enabling frequency reuse, enhances decoupling when the elements of the same MIMO antenna are detuned.

These two properties significantly improve the performance of the MIMO antenna. The key performance metrics of a MIMO system, including diversity parameters, are primarily determined by the gain and the level of coupling between the ports [19].

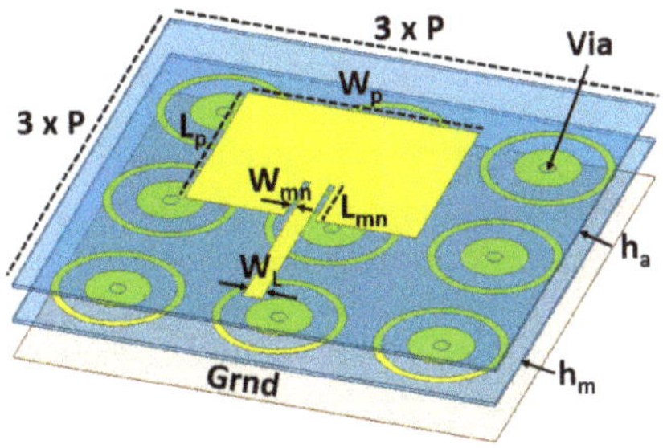

Figure 1. Schematic view of the proposed MIMO antenna structure based on a conventional patch antenna positioned above a metasurface.

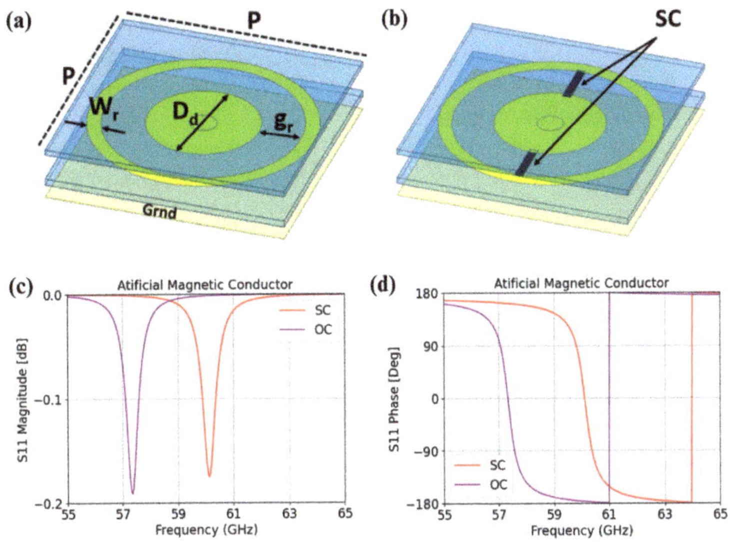

Figure 2. Schematic view of the metasurface unit cell (**a**) in OC configuration and (**b**) in SC configuration, with dielectric layers drawn with a separation between them and a simulation of the AMC behavior of the metasurface with (**c**) the magnitude of the reflected E field and (**d**) phase of the reflected E field at the patch level.

To ensure polarization insensitivity, each elementary cell of the metasurface is designed circularly. These cells are arranged in a square grid to maintain symmetry with the patch antenna, as depicted in Figure 1, and to the side-to-side symmetry of the MIMO antenna array described later. The elementary cells consist of metallic disks that are short-circuited at the center to the ground plane through a via. As shown in Figure 2, a metallic ring is incorporated around each disk to modulate the electromagnetic response. The frequency resonance agility is achieved by tuning the spacing or the coupling between the disk and the surrounding ring. This can be electronically controlled by adding diodes between the disk and the ring to alter the capacitive coupling without changing the physical dimensions. Each diode will exhibit capacitive coupling behavior, acting as an open circuit (OC) or a short circuit (SC) depending on the applied DC bias. Although the current prototype does not include diodes due to the challenges of integration and of biasing at this stage of development, their effects have been replicated.

In future prototypes including diodes, separated diodes can be picked and placed before the ground metal level for antenna, or diodes are fabricated in the carrier substrate.

Prior to the ground plane for antennas, first levels ensure bias interconnection. Opening through the ground plane can allow some via to reach the metasurface. A design must be adapted to include such a way to connect the disk to the ring.

Instead of diodes, a 40 µm wide metal strip is used (two strips on each side aligned with the patch antennas), allowing the resonant frequency of the antenna to be effectively varied by simulating the OC (Figure 2a) or SC behavior (Figure 2b) of the diodes.

Simulations of metasurface unit cells and of the full array, followed by optimizations with the patch antennas, led to the final design, which had a disk diameter, D_d, of 500 µm; a ring width, W_r, of 62.7 µm; and a separation, g_r, of 189.6 µm between the disk and the ring. The diameter via D_v is 120 µm. The metasurface's periodicity is p with 1.256 mm. A manufacturing constraint is not to exceed 30 µm in thickness for each BCB layer.

The patch antennas had dimensions of L_p = 1.8 mm × W_p = 1.5 mm. These were fed by 50 Ω microstrip lines 150 µm wide (W_L), with impedance matching notches of L_{mn} = 0.48 mm × W_{mn} = 0.044 mm. A perspective view of two antennas is shown in Figure 3a. The metasurface was placed h_m = 30 µm above the ground plane, with a h_a = 25 µm BCB layer separating it from the patch antennas. The patch was surrounded by nine unit cells of the metasurface. In Figure 3a, the MIMO antenna on the left left is in the OC configuration, while the antenna on the right is in the SC configuration. Simulation of S-parameters between the two ports and radiation patterns are shown in Figure 3b,c.

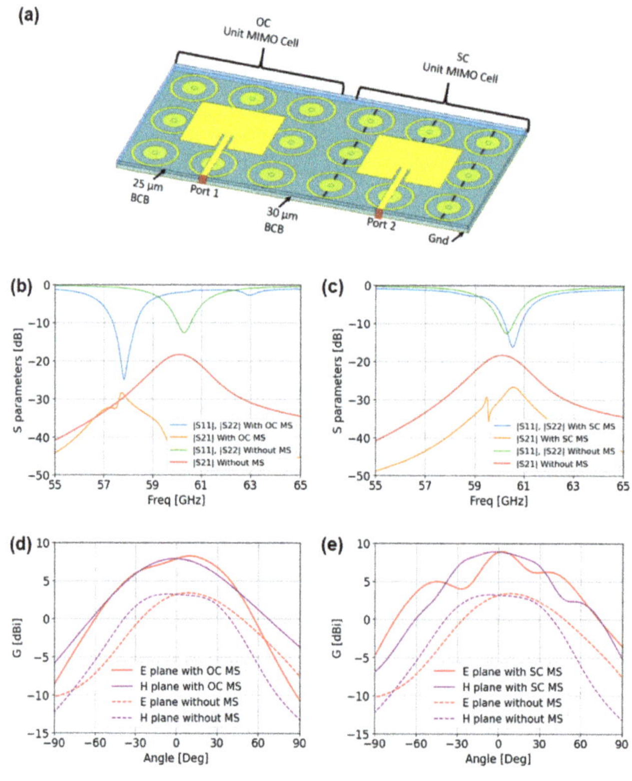

Figure 3. (**a**) Schematic view of the 2 × 2 MIMO antenna, showing the OC and SC configurations. (**b**) S-parameter comparison with and without the metasurface (MS) in the OC configuration. (**c**) S-parameter comparison with and without the MS in the SC configuration. (**d**) Simulated radiation patterns at 57.75 GHz for the OC configuration, with and without the MS. (**e**) Simulated radiation patterns at 60.5 GHz for the SC configuration, with and without the MS.

We observed in Figure 3b that the antenna in the open-circuit configuration achieves a bandwidth of 1000 MHz, between 57.25 GHz and 58.25 GHz, with a gain of 7.2 dBi to 8.6 dBi and an efficiency of 90%. The antenna in a short-circuited configuration offers a bandwidth of 700 MHz, ranging from 60.2 GHz to 60.9 GHz. This configuration provides a gain that varies between 8.5 dBi and 9 dBi with an efficiency of 90%. The isolation seen with S21 is better than −35 dB. To highlight the effects of the metasurface, Figure 3 presents S-parameters and the realized gain of a 2 × 2 MIMO antenna with 55 µm of BCB without the metasurface (MS) and with the MS in its two configurations. The results clearly show that the coupling is significantly reduced by approximately −10 dB when the MS is introduced. Furthermore, the realized gain is enhanced by around 4 dB. These observations confirm the beneficial impact of the metasurface in mitigating coupling and improving the radiation performance of the antenna system without requiring additional spacing between antennas to reduce coupling or increased thickness to enhance the gain. Such enhancements demonstrate the potential of the metasurface as a valuable design element for optimizing antenna characteristics, particularly in scenarios where high isolation and improved gain are critical.

3. Diversity Parameters

When dealing with MIMO antennas, it is essential to evaluate their diversity parameters, particularly the envelope correlation coefficient (ECC) and diversity gain (DG) [20]. ECC measures the correlation between MIMO antenna ports and is a critical parameter in determining the level of isolation and independence between antenna elements. It can be calculated from the radiation patterns or S-parameters [20]. A low ECC value indicates reduced coupling and improved performance, as it ensures the signals received or transmitted by the MIMO elements are largely uncorrelated, and the ECC should be less than 0.5 [20]. Diversity gain, on the other hand, quantifies the improvement in signal reliability provided by multiple antennas under fading conditions, directly impacting the MIMO system's effectiveness in multipath environments. The DG should be equal to 10 dB [20]. The ECC and DG of the 2 × 2 MIMO antenna depicted in Figure 3a are shown in Figure 4a,b, respectively. The ECC is close to 0 and DG is close to 10 dB. The values of the ECC and DG are in the acceptable range and hence, the proposed design is suitable for different MIMO applications.

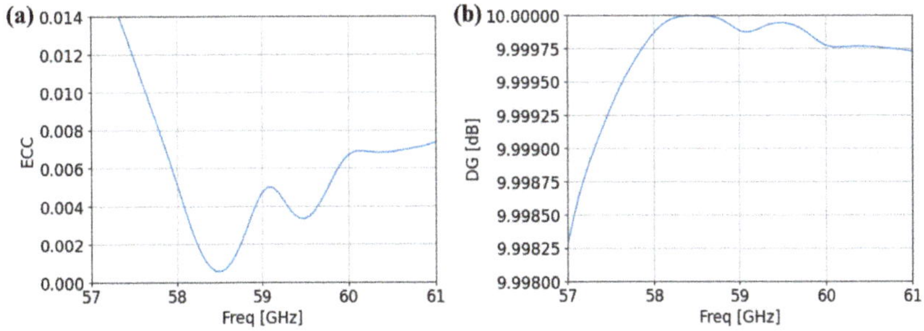

Figure 4. Diversity parameters in OC and SC configurations (**a**) for ECC and (**b**) for DG.

4. Fabrication and Experimental Validation

Using these OC and SC metasurface configurations, three different 4 × 4 MIMO structures were designed as shown in Figure 5 and fabricated to serve as proofs of concept for reconfigurability.

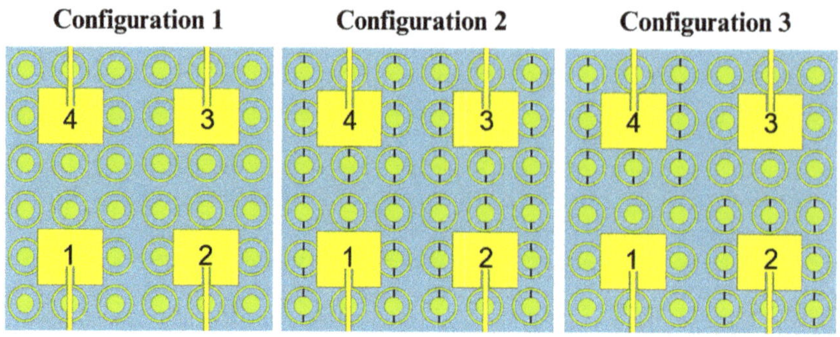

Figure 5. Schematic view of the 4 × 4 MIMO antenna, illustrating three OC and SC configurations. In Configuration 1, all four antennas are open circuit (OC); in Configuration 2, they are all short circuit (SC); and in Configuration 3, antennas 1 and 3 are OC, while antennas 2 and 4 are SC. Small black lines indicate short circuits.

The fabrication of the sample is achieved with a microelectronic manufacturing process. The manufacturing steps, with four UV lithographic levels, are illustrated in Figure 6. First, starting from a 500 µm thick two-inch silicon wafer, which serves as a carrier substrate, the ground plane is formed with a 1 µm thick Au layer, deposited by electron beam evaporation, with a 10 nm thick Ti adhesion layer. Next, interconnecting vias between the ground plane and the metal layer defining the metasurface are obtained by electrolytic growth through a 40 µm thick UV-sensitive resist layer (AZ40XT). After the growth of vias a little higher than 30 µm, the resist is replaced by BCB (Cyclotene3000-63 from DuPont, Wilmington, DE, USA) with a thickness of 30 µm and cured at 75%. A reactive-ion etching of BCB is performed through a photoresist mask to remove the residual BCB on top of the vias. This is the same mask used for the vias. A second 1 µm gold layer is deposited on the first BCB layer, on which a second electrolytic growth is performed for interconnecting vias up to the last third metal layer on top of the stack. Once the second level of vias is completed and the photoresist removed, the gold layer on top of the first BCB layer is etched by ion beam etching (Ar) to form the disks and rings of the metasurface. A second layer 25 µm thick of BCB is deposited and cured with the same parameters as the first BCB layer. It follows the etching of BCB on top of the vias, and a final 1 µm thick gold layer is deposited and etched to form the patch antennas, the microstrip lines, and the coplanar access pads for measurement with GSG probes.

Figure 7 shows the picture of the sample around the four combinations of MIMO antennas. For GSG probe measurement motivations with two independent GSG probes face-to-face or with a 90° orientation between them, 90° microstrip bends of the fed lines are used for some of the antennas. Each fed line includes the coplanar-to-microstrip transition. Some lines and reflects are included in the middle of the sample for calibration or de-embedding purposes.

To experimentally validate the fabricated structure, S-parameter and radiation pattern measurements are conducted using the setup illustrated in Figure 8. Around a probe station, this setup includes a two-port Keysight E8361C vector network analyzer (manufactured by Keysight Technologies, located in Santa Rosa, CA, USA), two GSG probes (100 µm pitch) for measuring reflection and coupling coefficients from 10 MHz to 67.5 GHz (uncalibrated up to 70 GHz), and a rectangular horn antenna (20 dB gain, WR-15 to coaxial 1.85 mm adapter) on a motorized arm for radiation pattern measurements. TRL calibration patterns are also manufactured in the middle of the sample (Figure 7).

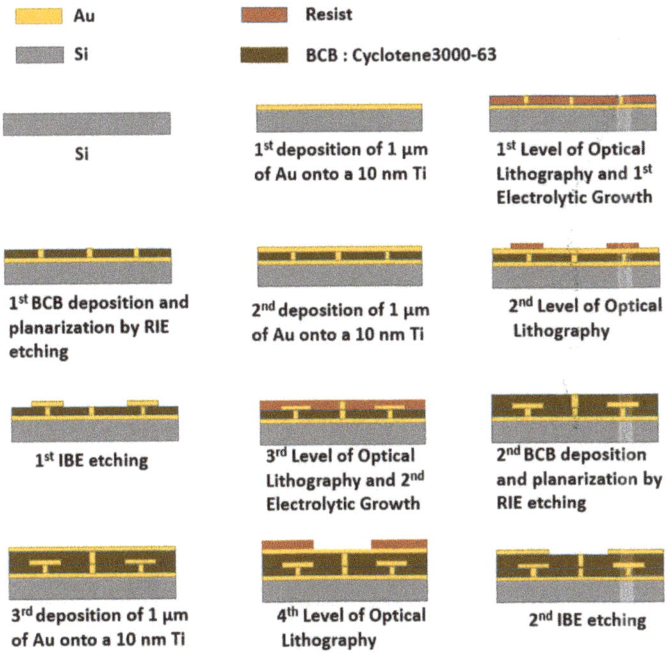

Figure 6. Illustration of the sample fabrication steps.

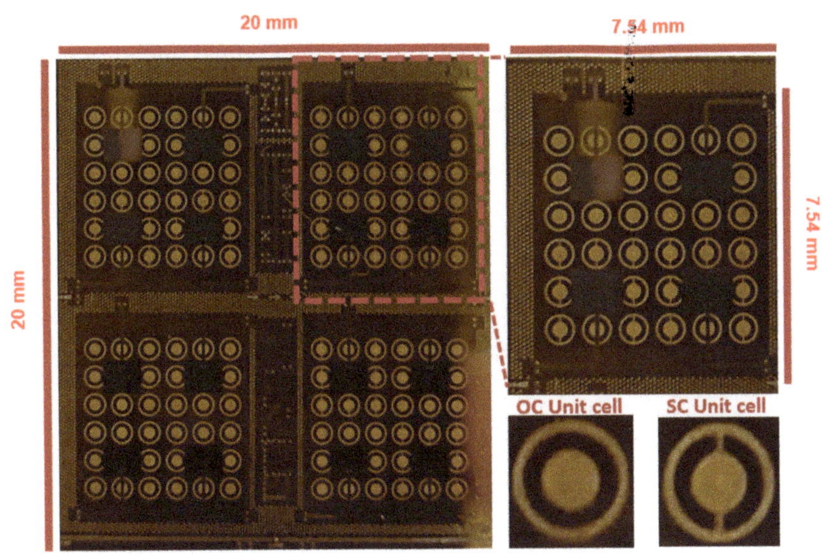

Figure 7. Picture of the fabricated prototype showing four zones with different 4 × 4 MIMO configurations, one 4 × 4 MIMO antenna magnified view, and open circuit unit cell and short-circuited unit cell magnified view.

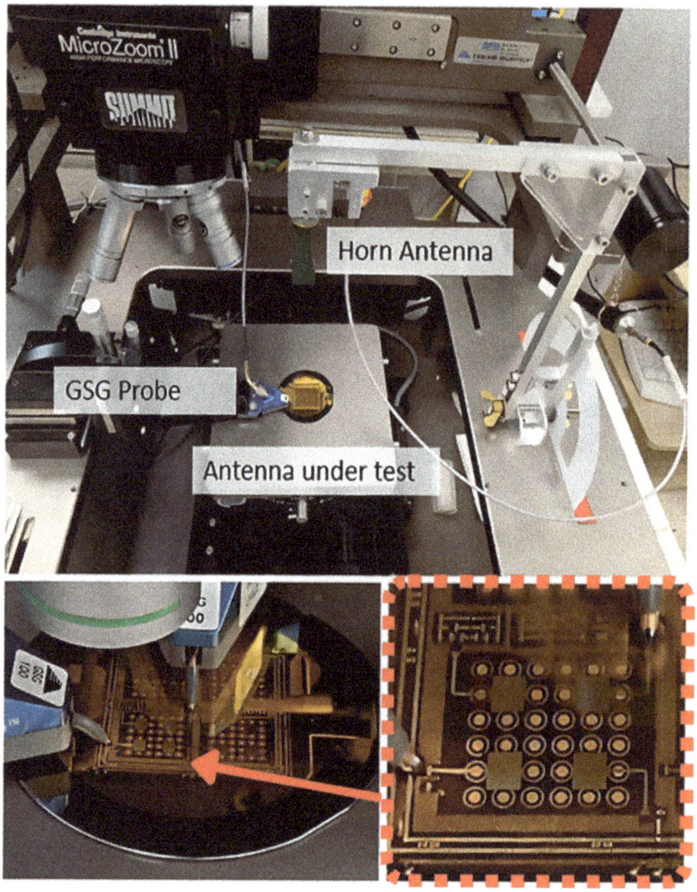

Figure 8. Fabricated sample and measurement setups of the scattering parameters and radiation patterns.

Measuring at 60 GHz presents challenges such as probe alignment sensitivity, calibration accuracy, and increased signal losses. The precise positioning of the GSG probes is crucial to ensure consistent contact, as even small misalignments can impact results. Additionally, the high atmospheric attenuation at 60 GHz requires careful control of the measurement distance to minimize losses while ensuring far-field conditions. Proper calibration and de-embedding are essential to obtain accurate measurements of the antenna's performance. A SOLT calibration is performed at the GSG probe level, ensuring good measurement for reflection and coupling coefficients. For radiation measurement, only relative transmission coefficients are obtained from planar antennas. The horn is connected to port 2 of the VNA, while the calibrated port 1 feeds the antenna with the same GSG probe. It should be noted that the metallic sample holder of the probe station acts as an infinitively wide ground plane. Antennas are on their two-inch-wide ground plane, which limits effective angular range.

5. Measurements Results

The measured and simulated S-parameters for these three structures are shown in Figure 9a–c, corresponding to configurations 1, 2, and 3, respectively. The antenna with an open-circuit metasurface configuration achieves a bandwidth of 1110 MHz, ranging from 56.45 GHz to 57.56 GHz. In contrast, the antenna with a short-circuited metasurface configuration offers a bandwidth of 1020 MHz, ranging from 59.96 GHz to 60.98 GHz. The antennas in configuration 3 achieve dual bandwidth: the first ranges from 57.2 GHz to 58.36 GHz, and the second is between 60.11 GHz and 60.99 GHz. As shown in Figure 9, the coupling observed with S41, S42, or S43 is lower than −35 dB for all configurations.

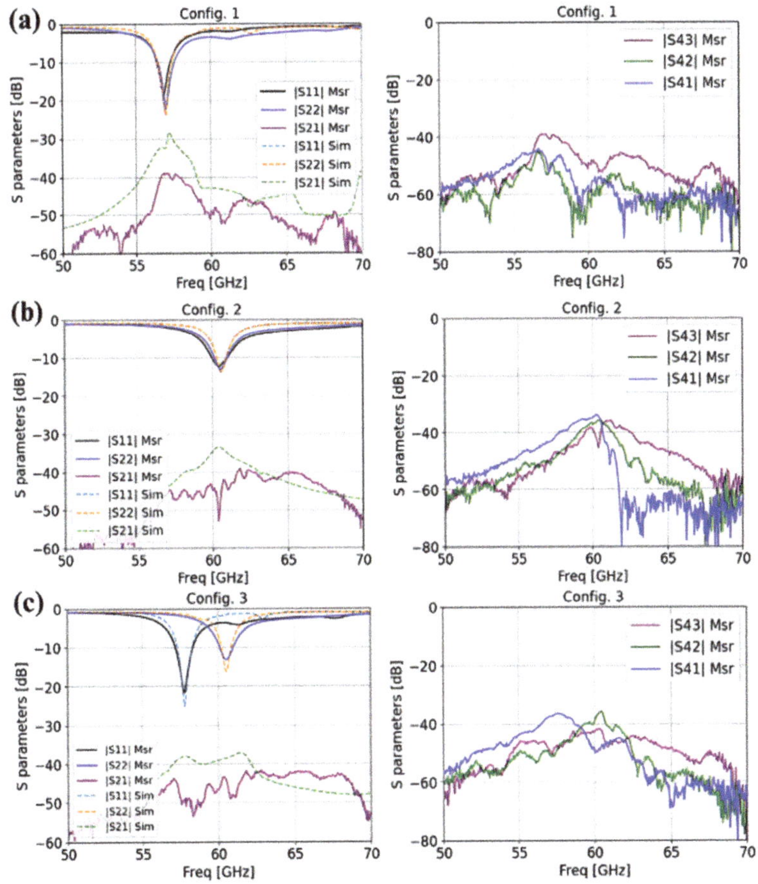

Figure 9. Measured and simulated reflection coefficients and coupling parameters for (**a**) configuration 1, (**b**) configuration 2, and (**c**) configuration 3.

The measured radiation patterns of the two configurations (OC and SC) in the E plane and H plane are shown in Figure 10. The measurements are in good agreement with the simulations. When comparing the simulations with the measurements, slight shifts can be observed. This discrepancy is due to the non-planarity of the BCB surface, caused by the presence of disks, rings, and vias at the metasurface interface, as well as some uncertainties in the exact values of permittivity and dielectric losses. It is also important to emphasize that the fabrication processes employed are not standardized, matured, or industrialized, as is the case for PCB technology. Instead, these processes have been meticulously adapted through iterative and labor-intensive procedures to achieve the current results, which are already promising. However, with further optimization and industrialization of the fabrication processes, we anticipate achieving significantly improved performance beyond the current outcomes.

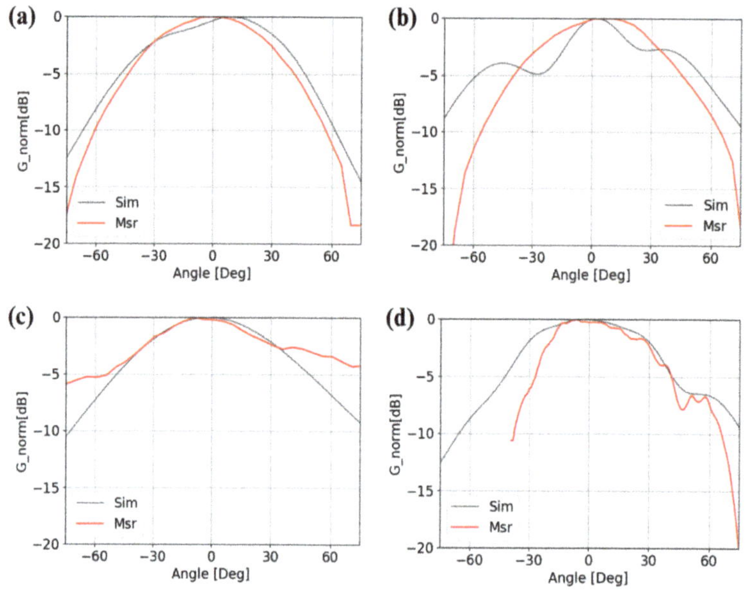

Figure 10. Radiation patterns. (**a**) E plane at 57.25 GHz in the OC metasurface configuration and (**b**) E plane at 60.5 GHz in the SC metasurface configuration; (**c**) H plane at 57.25 GHz in the OC metasurface configuration and (**d**) H plane at 60.5 GHz in the SC metasurface configuration.

6. Discussion

Several studies have demonstrated the potential of metasurface-based antennas to operate around 60 GHz, especially for WiGig and other high-data-rate applications. To this end, we have identified and listed several antennas operating around the 60 GHz frequency in order to compare their performance with that achieved by the proposed antenna, as summarized in Table 1. S. Ullah et al. [14] developed a 4 × 4 MIMO antenna on a 0.1 mm thick PCB substrate (Rogers, $\varepsilon r \approx 2.9$). The metasurface design, built with unit cells of 1.3 mm × 1.8 mm × 0.1 mm, allows for a wide frequency range of 57 to 63 GHz. This design achieved a gain of 8.2 dBi with good isolation levels reaching −54 dB. However, the size of the structure remains relatively large with 13 mm × 14 mm × 0.1 mm. Alassawi et al. [21] proposed both 2 × 2 and 4 × 4 MIMO-UWB antennas based on elliptic ring structures for 5G applications. Fabricated on Rogers 4003 ($\varepsilon r \approx 3.55$, thickness = 0.203 mm), the designs demonstrated good mutual coupling reduction through orthogonally arranged radiators. For the 2 × 2 configuration, the coupling achieved is −44 dB, while the 4 × 4 configuration reached −28 dB with the use of metallic isolators. Gains of 9.6 dBi (2 × 2) and 9.2 dBi (4 × 4) were achieved, demonstrating high efficiency for 5G communication systems. However, the dimensions of the 4 × 4 design remain significant (30 mm × 30 mm × 0.203 mm). Sharma et al. [22] introduced a compact 4 × 4 MIMO antenna operating at 60 GHz with dimensions of 16 × 16 × 0.254 mm^3. This structure achieved a gain of 10.56 dBi while maintaining isolation better than −50 dB, making it highly competitive for mmWave applications. However, the design does not yet include reconfigurability, which limits its flexibility for adaptive systems. Moreover, the dimensions are sufficiently large to consider their integration into an RFIC chip. Furthermore, the technology used, namely PCB printing, implies that the antenna will be used outside the chip, leading to a greater number of interconnections, which will further result in additional losses and performance degradation.

Table 1. Comparative study of the designed MIMO antenna with existing structures in the literature for 60 GHz applications.

Study	Technology	Dimensions (mm^3)	Freq. (GHz)	Gain (dBi)	Isolation (dB)	Reconfigurability
[14]	PCB (Rogers, $\varepsilon r \approx 2.9$)	$13 \times 14 \times 0.1$ ($2.6\lambda \times 2.8\lambda \times 0.02\lambda$)	57.0–63.0	8.2	−54	No
[21]	PCB (Rogers 4003, $\varepsilon r \approx 3.55$)	$30 \times 30 \times 0.203$ ($6.0\lambda \times 6.0\lambda \times 0.04\lambda$)	58.0–63.0	9.6 (2 × 2), 9.2 (4 × 4)	−44 (2 × 2), −28 (4 × 4)	No
[22]	Rogers RT/Duroid 5880, $\varepsilon r = 2.2$, h = 0.254 mm	$16 \times 16 \times 0.254$ ($3.2\lambda \times 3.2\lambda \times 0.05\lambda$)	58.925–60.66	10.56	<−50	No
Our work	On-chip, BCB substrate	$7.54 \times 7.54 \times 0.055$ ($1.51\lambda \times 1.51\lambda \times 0.01\lambda$)	57.0–63.0	6.5–9	<−35	Yes (proof of concept, OC/SC configurations)

The on-chip four-port MIMO antenna proposed here, including the advantages of a metasurface and of a BCB substrate, allows for achieving a more compact form factor and competitive performance. The overall size of our antenna is $7.54 \times 7.54 \times 0.055$ mm^3, which is significantly smaller than the aforementioned designs. Isolation between MIMO ports is maintained to be better than −35 dB, and gains ranging from 6.5 to 9 dBi are achieved depending on the operating frequency and on the metasurface configuration. Unlike previous works, our antenna is designed for frequency agility, with proof of concept presented here. Moreover, it should be noted that the integration of metasurfaces within this architecture further enhances the overall system performance by leveraging their unique properties for electromagnetic wave control. The roles of the metasurfaces used are well understood, but the main innovation does not lie in their application. Instead, we emphasize the complete integration of these technologies operating in the millimeter-wave band into an on-chip structure. This level of integration, achieved using cleanroom fabrication technologies commonly used in microelectronics, sets our work apart from most existing solutions in the literature, which primarily rely on PCB technologies. This approach reduces interconnections and thus losses, which constitutes a notable advancement. The future integration of diodes will enable dynamic reconfigurability, addressing a key limitation in existing designs.

This comparison demonstrates that our on-chip solution offers significant advantages in terms of miniaturization and reconfigurability while achieving competitive isolation and gain for mmWave applications.

7. Conclusions

In this work, we presented a 4 × 4 MIMO antenna integrated on a BenzoCycloButene (BCB) substrate and enhanced with a metasurface for high-performance operation at 60 GHz. With a compact volume of only 3.13 mm^3, the antenna delivers an impressive gain of 8 dBi, with an average efficiency of 90%, and the frequency tuning range extends from 56.5 GHz to 61 GHz by simply reconfiguring the metasurface. The innovative manufacturing process enables seamless integration over active circuitry and provides the option to incorporate diodes for dynamic frequency matching. This design approach not only enables significant miniaturization but also opens the door to scaling down the technology for submillimeter wave applications, making it highly suitable for next-generation wireless systems, such as communication at 120 GHz and beyond.

Author Contributions: Conceptualization, K.B., N.Z. and B.E.R.; Methodology, K.B., N.Z. and B.E.R.; Validation, K.B. and N.Z.; Writing—original draft, K.B. and B.E.R.; Writing—review & editing, N.Z. and B.E.R. All authors have read and agreed to the published version of the manuscript.

Funding: This research received no external funding.

Institutional Review Board Statement: Not applicable.

Informed Consent Statement: Not applicable.

Data Availability Statement: Data is contained within the article.

Acknowledgments: The samples were fabricated and measured at the Center for Nanosciences and Nanotechnologies (C2N–UMR9001). This work was partially supported by the French RENATECH network of technological facilities.

Conflicts of Interest: The authors declare no conflicts of interest.

Abbreviations

The following abbreviations are used in this manuscript:

BCB	BenzoCycloButene
MIMO	Multiple Input Multiple Output
AMC	Artificial Magnetic Conductor
SC	Short Circuit
OC	Open Circuit
ECC	Envelope Correlation Coefficient
DG	Diversity Gain
GSG	Ground–Signal–Ground

References

1. Statista. Volume of Data/Information Created, Captured, Copied, and Consumed Worldwide from 2010 to 2023, with Forecasts from 2024 to 2028. Available online: https://www.statista.com/statistics/871513/worldwide-data-created/ (accessed on 15 April 2022).
2. Wang, Y.; Li, J.; Huang, L.; Jing, Y.; Georgakopoulos, A.; Demestichas, P. 5G Mobile: Spectrum Broadening to Higher-Frequency Bands to Support High Data Rates. *IEEE Veh. Technol. Mag.* **2014**, *9*, 39–46. [CrossRef]
3. Daraseliya, A.; Korshykov, M.; Sopin, E.; Moltchanov, D.; Andreev, S.; Samouylov, K. Coexistence Analysis of 5G NR Unlicensed and WiGig in Millimeter-Wave Spectrum. *IEEE Trans. Veh. Technol.* **2021**, *70*, 11721–11735. [CrossRef]
4. Rappaport, T.; Xing, Y.; MacCartney, G.R.; Molisch, A.F.; Mellios, E.; Zhang, J. Overview of millimeter wave communications for fifth-generation (5G) wireless networks—with a focus on propagation models. *IEEE Trans. Antennas Propag.* **2017**, *65*, 6213–6230. [CrossRef]
5. Alam, M.S.; Siddiqui, S.T.; Qidwai, K.A.; Aftab, A.; Kamal, M.S.; Shahi, F.I. Evolution of Wireless Communication Networks from 5G to 6G: Future Perspective. *Radioelectron. Commun. Syst.* **2023**, *66*, 213–222. [CrossRef]
6. FCC. FCC Opens Spectrum Horizons for New Services and Technologies. Available online: www.fcc.gov/document/fcc-opens-spectrum-horizons-new-services-technologies (accessed on 30 October 2022).
7. Xiao, M.; Mumtaz, S.; Huang, Y.; Dai, L.; Li, Y.; Matthaiou, M.; Karagiannidis, G.K.; Bjornson, E.; Yang, K.; I, C.-L.; et al. Millimeter wave communications for future mobile networks. *IEEE J. Sel. Areas Commun.* **2017**, *35*, 1909–1935. [CrossRef]
8. Heydari, P. Terahertz Integrated Circuits and Systems for High-Speed Wireless Communications: Challenges and Design Perspectives. *IEEE Open J. Solid-State Circuits Soc.* **2021**, *1*, 18–36. [CrossRef]
9. Alibakhshikenari, M.; Ali, E.M.; Soruri, M.; Dalarsson, M.; Naser-Moghadasi, M.; Virdee, B.S.; Stefanovic, C.; Pietrenko-Dabrowska, A.; Koziel, S.; Szczepanski, S.; et al. A comprehensive survey on antennas on-chip based on metamaterial, metasurface, and substrate integrated waveguide principles for millimeter-waves and terahertz integrated circuits and systems. *IEEE Access* **2022**, *10*, 3668–3692. [CrossRef]
10. Ghosh, S.; Sen, D. An inclusive survey on array antenna design for millimeter-wave communications. *IEEE Access* **2019**, *7*, 83137–83161. [CrossRef]
11. Karim, R.; Iftikhar, A.; Ramzan, R. Performance-Issues-Mitigation-Techniques for On-Chip-Antennas—Recent Developments in RF, mm-Wave, and THz Bands with Future Directions. *IEEE Access* **2020**, *8*, 219577–219610. [CrossRef]
12. Pendry, J.B.; Holden, A.J.; Stewart, W.J.; Youngs, I. Extremely low frequency plasmons in metallic mesostructures. *Phys. Rev. Lett.* **1996**, *76*, 4773. [CrossRef] [PubMed]
13. Holloway, C.L.; Kuester, E.F.; Gordon, J.A.; O'Hara, J.; Booth, J.; Smith, D.R. An overview of the theory and applications of metasurfaces: The two-dimensional equivalents of metamaterials. *IEEE Antennas Propag. Mag.* **2012**, *54*, 10–35. [CrossRef]

14. Ullah, S.; Yeo, W.-H.; Kim, H.; Yoo, H. Development of 60-GHz millimeter wave, electromagnetic bandgap ground planes for multiple-input multiple-output antenna applications. *Sci. Rep.* **2020**, *10*, 8541. [CrossRef] [PubMed]
15. DuPont. Cyclotene 3000 Series. Technical Data Sheet. 2024. Available online: https://kayakuam.com/wp-content/uploads/2024/05/888_00006_DuPont_Cyclotene_3000_Series.pdf (accessed on 30 October 2024).
16. Perret, E.; Zerounian, N.; David, S.; Aniel, F. Complex permittivity characterization of benzocyclobutene for terahertz applications. *Microelectron. Eng.* **2008**, *85*, 2276–2281. [CrossRef]
17. Zerounian, N.; Aouimeur, W.; Grimault-Jacquin, A.-S.; Ducournau, G.; Knightly, E.W. Coplanar waveguides on BCB measured up to 760 GHz. *J. Electromagn. Waves Appl.* **2021**, *35*, 2051–2061. [CrossRef]
18. Ghasempour, Y.; Da Silva, C.R.C.M.; Cordeiro, C.; Knightly, E.W. IEEE 802.11 ay: Next-generation 60 GHz communication for 100 Gb/s Wi-Fi. *IEEE Commun. Mag.* **2017**, *55*, 186–192. [CrossRef]
19. Mane, P.R.; Kumar, P.; Ali, T.; Alsath, M.G.N. Planar MIMO antenna for mmWave applications: Evolution, present status & future scope. *Heliyon* **2023**, *9*, 2.
20. Sharma, P.; Tiwari, R.N.; Singh, P.; Kumar, P.; Kanaujia, B.K. MIMO antennas: Design approaches, techniques and applications. *Sensors* **2022**, *22*, 7813. [CrossRef] [PubMed]
21. Alassawi, S.A.; Ali, W.A.E.; Ismail, N.; Rizk, M.R.M. Compact elliptic ring 2×2 and 4×4 MIMO-UWB antenna at 60 GHz for 5G mobile communications applications. *Microsyst. Technol.* **2023**, *29*, 431–440. [CrossRef]
22. Sharma, M.; Kumar, A.; Kikan, V.; Jaitly, G.; Bhardwaj, S.; Neha; Bano, T. Conformal ultra-compact narrowband 60.0 GHz four-port millimeter wave MIMO antenna for wearable short-range 5G application. *Wirel. Netw.* **2024**, *30*, 1815–1831. [CrossRef]

sensors

MDPI

Article

Dual-Band Filter and Diplexer Design Using Extremely Miniaturized Substrate-Integrated Coaxial Cavity

Chun-Ming Hung [1], Ci-Fang Jheng [2], Keh-Yi Lee [3], Chung-I G. Hsu [4] and Min-Hua Ho [2,*]

1 Universal Microwave Technology Inc., Keelung 206, Taiwan; alan_hung@umt-tw.com
2 Department of Electronic Engineering, National Changhua University of Education, Changhua 500, Taiwan; m1253004@mail.ncue.edu.tw
3 Department of Electrical Engineering, Chinese Culture University, Taipei 111, Taiwan; kyleentu@ms52.hinet.net
4 Department of Electrical Engineering, National Yunlin University of Science and Technology, Yunlin 640, Taiwan; cghsu@yuntech.edu.tw
* Correspondence: ho@cc.neue.edu.tw; Tel.: +886-47128241

Highlights:
We utilize printed circuit board technology to develop a substantially miniaturized substrate-integrated coaxial cavity (SICC) resonator, achieving a significant size reduction as the circuit occupies a mere 2.1% of the area of its substrate-integrated waveguide cavity counterpart when operating at the same frequency. For the first time, these miniaturized SICC resonators have been successfully incorporated into the design of a dual-band filter and a diplexer, demonstrating a considerable performance improvement.

Abstract: This paper presents the design of a dual-band filter and a diplexer using an extremely miniaturized substrate-integrated coaxial cavity (SICC) structure. The presented dual-band filter can function as a front-end circuit block connected to 5G antennae, enabling dual-passband operation for 5G applications. The diplexer is designed for use in 5G communication systems, positioned after the 5G antennae to facilitate the switching of transmitting (Tx) and receiving (Rx) signals between the Tx and Rx terminals. The main contribution of this work is the development of a highly miniaturized substrate-integrated coaxial cavity (SICC) to design a dual-band filter (DBF) and a diplexer. The circuit area of the proposed dual-frequency SICC is a mere 2.1% of its conventional substrate-integrated waveguide (SIW) cavity counterpart when operating at the same frequency. A dual-band filter and a diplexer are realized using two and three highly miniaturized SICC resonators, respectively. The dual-band filter is designed to have a transmission zero on each passband side to enhance signal selectively. At most in-band frequencies, the isolation between the diplexer's channel bands exceeds 20 dB. A sample dual-band filter and diplexer have been fabricated for experimental validation, demonstrating excellent agreement between the measured and simulated data. To the best of the authors' knowledge, the designed dual-band filter and diplexer achieve the highest circuit area efficiency within the categories of dual-band SIW cavity filters and diplexers.

Keywords: dual-band filter (DBF); diplexer; substrate-integrated waveguide (SIW); substrate-integrated coaxial cavity (SICC); circuit miniaturization

1. Introduction

Dual-band filters and diplexers that utilize substrate-integrated waveguide (SIW) technology have attracted the attention of many researchers. The employed SIW structure was first introduced by Zaki et al. [1]. Subsequently, Fuji et al. [2] and Deslandes and

Wu [3] reported similar structures under different nomenclatures. The SIW possesses several advantages, including a higher quality factor and enhanced noise immunity, which are not present in other planar structures such as microstrip lines, coplanar waveguides, and slotlines. However, the size of SIW filters is considered excessively large for modern communication systems, particularly when operating at half-wavelength frequencies. Numerous studies have reported various methodologies aimed at reducing the circuit size of SIWs [4–20]. Notably, references [4–6] utilize half-mode and folded SIW structures, respectively, reducing the circuit area by a factor of two. Introducing the quarter-mode SIW cavity in [7–11] and the eighth-mode SIW cavity in [10–12] necessitates only a quarter/one-eighth of the size of a conventional SIW cavity. In [13–15], the size miniaturization of the SIW BPF is accomplished by incorporating complementary split-ring resonators (CSRRs), facilitating the dominant propagation mode to operate at a frequency much lower than the cut-off frequency, thereby effectively decreasing the filter size. However, the filters are susceptible to radiation losses due to the open edges of the SIW and the slots of the CSRR, limiting their applicability to low microwave frequency ranges. In [16], inserting a dielectric rod within the SIW cavity excites the TM_{nm0} mode, resulting in a 33% reduction in size. A more substantial size reduction of 60% is reported in [17] by implementing a concave–convex cavity filter. Recently, an exceptional miniaturization design was introduced in [18], achieving a 93% reduction in circuit size by loading the quarter-mode SIW cavity with a complicated capacitive structure. Furthermore, the evanescent-mode SIW cavities, also referred to as coaxial resonators or embedded coaxial SIWs, in [19–21] demonstrate even greater size reductions, with circuit area reductions exceeding 96% for a single SIW cavity resonator.

In recent years, bandpass filters (BPFs) with dual-band functions have received immense demand since wireless communication systems exhibit a trend toward a multichannel operation. Numerous substrate-integrated waveguide (SIW) BPFs [22–27] have been developed to meet the dual-passband requirements. Among these, a compact dual-band BPF was designed in [22], utilizing multiple SIW cavities arranged in a vertically stacked configuration. However, the slot incurs excessive losses when utilized to subdue common-mode signals at high frequencies. The SIW cavity BPF of dual-mode response presented in [23] utilizes two metallic rods of differing sizes, each capable of independently tuning its associated resonance frequency, thereby facilitating the formation of two independently controlled passbands. In [24], a single circular SIW cavity operating under multiple transverse magnetic (TM) modes is utilized for dual-band filter design. The independent control of the dual-band frequencies is ingeniously achieved by perturbing the TM modes using additional via-holes in conjunction with the slots embedded in the cavity wall. In [25], the dual-band response is realized by incorporating complementary split-ring resonators (CSRRs) embedded in the SIW cavity walls, leveraging degenerate-mode resonance. Each CSRR acts as an additional resonance node extending from the cavity, creating transmission zeros (TZs) on either side of passband edges. Furthermore, as noted in [26], an E-slot embedded in the SIW cavity renders the latter a dual-band function, with the resonances of the E-slot's even- and odd-mode resonances dominating the passbands' frequencies. Moreover, the half-mode rectangular SIW cavity was initially utilized in dual-band filter applications, as discussed in reference [27], allowing for a significant separation between the two passbands. Overall, the aforementioned SIW cavities in [22–27] are relatively large and, as such, are not well-suited for commercial microwave systems that necessitate circuit miniaturization.

The dual-band SIW cavity BPFs are also applicable in the design of diplexers, which facilitate the routing of signals at two distinct frequencies to separate ports. Recent studies have reported several innovative diplexers utilizing SIW cavity designs [28–33]. For instance, the hexagonal SIW cavity diplexer described in [28] employs cavities that operate under perturbed degenerate-mode resonance for the transmitting (Tx) and receiving (Rx) filters. However, a notable limitation of this design is the relatively large size of the cavities utilized. In [29], a diplexer is constructed using two cavities that operate under perturbed dual-mode resonances. Each cavity generates a transmission zero (TZ) positioned within the passband of the opposing cavity filter to enhance isolation. Although this design incorporates only two SIW cavities, it necessitates a substantial SIW T-junction to connect the common port to both the Tx and Rx cavities. The work presented in [30] introduces a sophisticated diplexer design that features frequency-tuning capabilities and accommodates various port configurations, including single-ended and balanced forms. Frequency tuning is achieved by incorporating piezoelectric disks into the cavity, which load the cavity with distinct capacitances. A compact SIW diplexer composed of combined single- and dual-mode cavities is documented in [31]. This configuration enables the two duplexing bands to have a flexible bandwidth ratio; however, it requires a considerable circuit area due to the necessity of multiple SIW cavities. In [32], integrating a microstrip filter within three air-filled circular SIW cavities successfully facilitates the separation of two duplexing bands. The microstrip filter routes the low-frequency signal in this configuration, while the air-filled SIW cavities channel the high-frequency millimeter-wave signal. Reference [33] uses a field perturbation on the SIW's multi-mode to create the filter's multiple passband functions. The aforementioned SIW multi-band BFSs and diplexers exhibit a relatively large circuit area, which may limit their suitability for commercial applications.

In our previous study, we investigated a miniaturized substrate-integrated coaxial cavity (SICC) structure, which has been effectively utilized in the designs of size-reduced single-band [20,34], dual-band [35], tri-band [36], and quad-band [37] BPFs, as well as a diplexer [35]. While SICC technology has achieved and demonstrated the highest circuit area efficiency for substrate-integrated waveguide (SIW)-related cavities, ongoing advancements in the reduction of SIW cavity size are still being pursued. Reference [38] illustrates a further size reduction technique for the SICC structure by enhancing the SICC's loading capacitance through a split coplanar waveguide ring (SCR). However, the current advancements pertain solely to single-band filters (BPFs). Consequently, there is a need for further research and study into dual-band BPFs and diplexers that utilize this ultra-miniaturized SICC technology. Hence, this paper presents the design of a dual-band BPF and diplexer employing the extremely size-reduced SICC structure. Sample circuits have been fabricated for experimental validation, revealing good agreement between the measured and simulated data.

2. Implementation of Miniaturized Substrate-Integrated Coaxial Cavity

The proposed extremely miniaturized substrate-integrated coaxial cavity (SICC) and its corresponding equivalent circuit model are illustrated in Figure 1. This SICC consists of two tightly stacked RT/Duroid 5880 substrates (ε_r = 2.2 and tanδ = 0.0009) with thicknesses h_1 = 0.254 mm for the top substrate and h_2 = 1.57 mm for the bottom substrate. Additionally, the configuration includes three metal sheets, each with a thickness of 35 μm. A 0.08 mm thick prepreg (PP) layer, with ε_r = 4 and tanδ = 0.013, is positioned between the middle metal layer and the top substrate to facilitate substrate binding. The equivalent circuit model depicted in Figure 1d consists of a short-circuited coaxial line, with a length of h_2, which is loaded with capacitances C_p, C_{SCR}, and an inductance L_{via}. The capacitance C_p, approximated by $C_p \approx 0.25\pi\varepsilon D^2(1/h_1 + 1/h_2)$, represents the combined capacitance between the circular patch of diameter D (in the M2 layer) and the top and bottom walls. The term C_{SCR} stands for the capacitance between the center strip of the embedded split coplanar waveguide ring (SCR) and the bottom ground plane. At the same time, L_{via} denotes the inductance associated with the three blind via-holes that connect the embedded SCR to the circular patch. It is noteworthy that, as referenced in [39], C_{SCR} can be evaluated as follows:

$$C_{\mathrm{SCR}} = \left(2\varepsilon_0(\varepsilon_r - 1)\frac{K(k_1)}{K(k')} + 4\varepsilon_0\frac{K(k_0)}{K(k_0')}\right)l_{\mathrm{SCR}} \tag{1a}$$

$$k_0 = S/(S+2W),\ k_0' = \sqrt{1-k_0^2} \tag{1b}$$

$$k_1 = \frac{\sinh(\pi S/4h_2)}{\sinh[\pi(S+2W)/4h_2]},\ k_1' = \sqrt{1-k_1^2}, \tag{1c}$$

where $K(\cdot)$ is the complete elliptic integral of the first kind, while the parameters S and W denote the strip and slot width of the embedded SCR, respectively. Moreover, ε_0 is the permittivity in vacuum, and ε_r is the substrate's dielectric constant. Additionally, l_{SCR} refers to the mean arc length of the SCR, measured in millimeters (mm). Based on the equivalent circuit model, the total susceptance $B(\omega)$ between the circular patch and the ground, as illustrated in Figure 1d, is expressed as follows:

$$B(\omega) = \omega\left(\frac{C_{\mathrm{SCR}}}{1-\omega^2 L_{\mathrm{via}}C_{\mathrm{SCR}}} + C_{\mathrm{p}}\right) - Y_0\cot\beta h_2 \tag{2}$$

where the phase constant is denoted by β, with c_0 representing the speed of light in vacuum and Y_0 indicating the characteristic admittance of the short-circuited coaxial transmission line formed by the blind via-ring fence of diameter D_v and the four thru-via side walls of the square cavity. For convenience, the frequency-dependent expression in the parentheses of Equation (2) is referred to as the effective loading capacitance C_{eff}. This expression is applicable when the operating frequency is much lower than the series resonance frequency of the L_{via}-C_{SCR} branch.

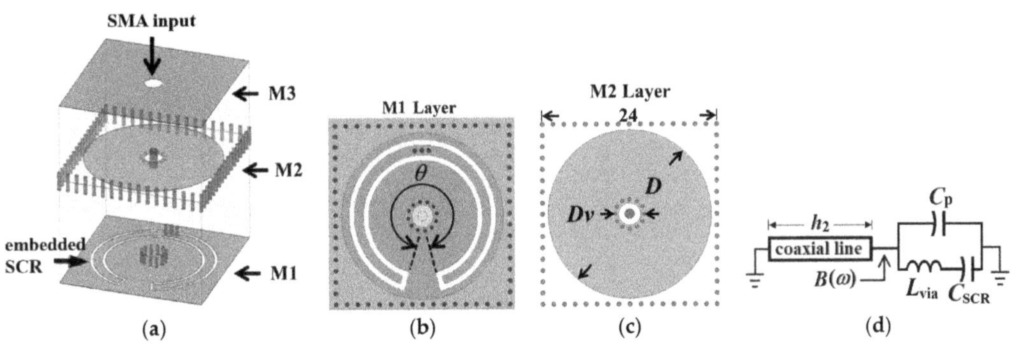

Figure 1. Layouts of the extremely miniaturized substrate-integrated coaxial cavity: (**a**) 3D view, (**b**) bottom (M1) layer embedded with the split CPW ring (with the subtended angle θ), (**c**) middle (M2) layer, and (**d**) the equivalent circuit model.

In a conventional square SIW cavity, the fundamental TE_{101} exhibits a resonance frequency denoted as f_0. In addition to this fundamental mode, the proposed SICC resonates as a capacitively loaded coaxial-line mode oriented along the normal direction of the substrate, characterized by a resonance frequency of f_r. Since the lower substrate has a small electric thickness βh_2 around the desired operating frequencies, the approximation $\cot(\beta h_2)\approx 1/\beta h_2$ can be made, causing the condition $B(\omega) = 0$ in Equation (2) to yield the resonance frequency f_r as [20].

$$f_r = \frac{1}{2\pi}\left[\frac{Y_0 c_0}{C_{\mathrm{eff}}\sqrt{\varepsilon_r}h_2}\right]^{1/2} \tag{3}$$

In contrast to our previous works in [34–37], which incorporated only a single capacitance within the cavity, the proposed SICC presented in this paper significantly enhances the overall capacitance. This enhancement is attributed to integrating an additional capacitance, which is facilitated by the SCR embedded in the M3 layer, as illustrated in Figure 1a. The supplementary capacitance can now be integrated with the capacitance C_p through three connection via-holes to form the enhanced effective loading capacitance C_{eff}. This configuration not only enhances the loading capacitance of the SICC but also leads to a further downshift in the resonance frequency. In this paper, the side dimension of the square SICC and the conventional square SIW cavity counterpart is fixed at 24 mm. The resonance frequency of the conventional square SIW cavity is f_0 = 5.96 GHz. The frequency ratios f_r/f_0 and C_{eff}, with their values calculated as functions of the SCR's subtended angle (θ) in the SICC, are omitted here and can be found in [38]. It should be noted that the SICC area is proportional to $(f_r/f_0)^2$. The outcome of $f_r/f_0 < 1$ indicates that the circuit size has been reduced, and all values of f_r/f_0 obtained in this paper are less than 0.15.

Figure 2 illustrates the normalized resonance frequency of the SICC depicted in Figure 1 as a function of the subtended angle (θ) of the embedded SCR. The figure presents the simulated frequency ratio (f_r/f_0) against the SCR's subtended angle (θ) for various diameters of the via-ring fence (D_v), with the outer diameter of the SCR fixed at 20.8 mm. The data indicate that an increase in the subtended angle (θ) corresponds to a decrease in the resonance frequency (f_r). This phenomenon can be attributed to the fact that a larger value of θ increases the capacitance C_{SCR} for the specified dimensions of the SCR. However, the drawback of the resonance frequency downshift is a decrease in the Q_u value, as seen in Figure 3. This is because the larger the value of θ, the larger the conduction and radiation losses due to the longer aperture in the cavity's bottom wall. Furthermore, because the blind via-ring fence in the cavity's lower substrate region emulates the coaxial line's center conductor, a smaller via-ring-fence diameter (D_v) yields a smaller characteristic admittance (Y_0), resulting in a lower f_r, as is evident from Figure 2 and Equation (3).

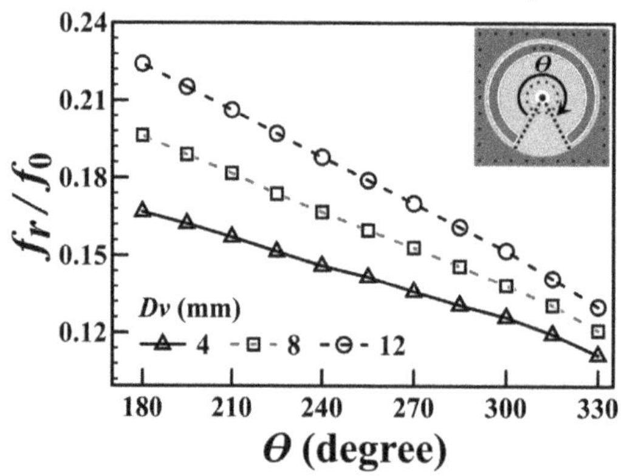

Figure 2. Normalized resonance frequency (f_r/f_0) versus the SCR's subtended angle (θ) for several via-ring-fence diameters (D_v).

As previously mentioned, an increase in the subtended angle leads to a decrease in the unloaded Q_u value. Figure 3a illustrates the relationship between the unloaded quality factor Q_u and the SCR's subtended angle. Additionally, Figure 3b presents the variation of Q_u as a function of the SCR's bilateral slot gap dimension (g), revealing that a reduced gap dimension leads to a smaller Q_u value. This phenomenon can be attributed to the greater conduction currents at the edges of the smaller gap, which result in increased conduction losses. Notably, Q_u is relatively insensitive to changes in the diameter (D_v) of the via-ring fence.

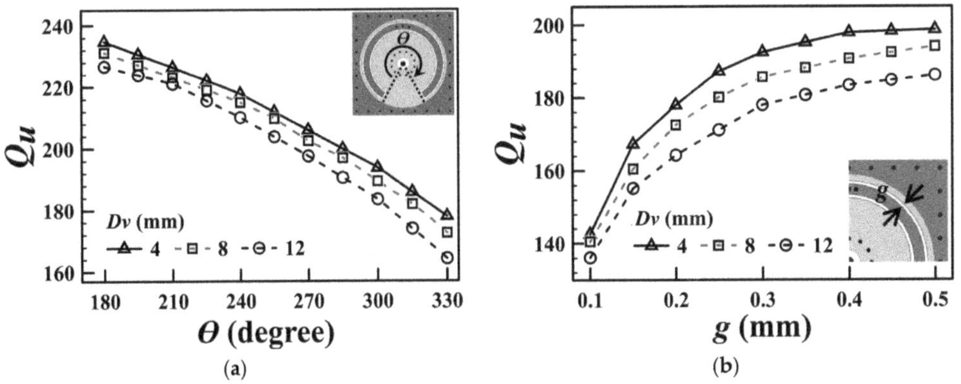

Figure 3. Simulated Q_u values as a function of (**a**) SCR's subtended angle (θ) and (**b**) the SCR's slot gap dimension (g) for several values of D_v.

3. Miniaturized Dual-Band SICC BPF

A sample miniaturized dual-band SIW cavity BPF is presented in this session. Figure 4 shows the structure of the proposed extremely miniaturized dual-band SICC BPF. The proposed dual-band BPF is composed of two SICCs (Figure 1) in a vertically stacked form, as shown in Figure 4a. Each cavity comprises two tightly bonded RT/Duroid 5880 substrates of 1.57 (top substrate) and 0.254 mm (bottom substrate) thicknesses. Notably, one cavity is inverted and positioned atop the other, resulting in a total of four substrate layers (five metal layers) for the filter. The feeding structure consists of a sub-miniature version A (SMA) connector on the cavity's top wall. The SMA probe extends through the cavity and connects to the bottom wall, establishing a short-circuit feed. Figure 5 gives the layouts of the three metal layers (see Figure 4b) and their circuit dimensions (in mm). The middle (M2) layer exhibits two semicircular patches, each of which is connected to the M1 layer of the cavity through one (for the right-side patch) or two extra blind via-holes (for the left-side patch). The former is associated with the low-frequency resonance, and the latter with the high-frequency resonance. In other words, one square cavity houses two resonators. For convenience, we can call each of the two resonators in the same cavity a half-mode coaxial resonator. The low-frequency resonance predominantly exists in the right half of the cavity, whereas the high-frequency one exists in the left half of the cavity. When the dimensions of the semicircular patch are fixed, the resonance frequency of each half-mode coaxial resonator is predominantly influenced by the dimensions of the associated SCR and the extra blind shorting via-holes positioned at appropriate locations. In Figure 1, the circular patch and the bottom cavity wall are connected using the blind via-ring fence of diameter D_V. This blind via-ring fence, in conjunction with the surrounding cavity, forms a short-circuited coaxial line. Under the condition that the lower substrate has a small electric thickness, the equivalent lumped element looking into the short-circuited coaxial line from the circular patch is an inductance. In the half-mode coaxial resonator, the extra blind shorting via-holes connect the semicircular patch to the bottom cavity wall, and they play the same role as the blind via-ring fence in Figure 1.

The equations presented in Section 2 may no longer accurately evaluate the resonance frequencies for the proposed structure of two resonators within one cavity. However, the trend of the resonance frequencies is similar to that of the complete SICC case. Nevertheless, an empirical approach combined with an L-C circuit model, as utilized in [35,36], is used to determine the resonance frequency of each half-mode coaxial resonator. This method involves fixing the loaded capacitances (i.e., fixing the size of the semicircular patch and the SCR) while varying the number and positions of the extra shorting vias that connect the semicircular patch to the bottom wall of the cavity, effectively controlling the inductance value for each L-C resonance. Intuitively, the right-side half-mode coaxial resonator with only one extra blind via-hole can be regarded as having a larger effective inductance in the L-C circuit model than the left-side one with two extra blind via-holes, causing the former to have a lower resonance frequency.

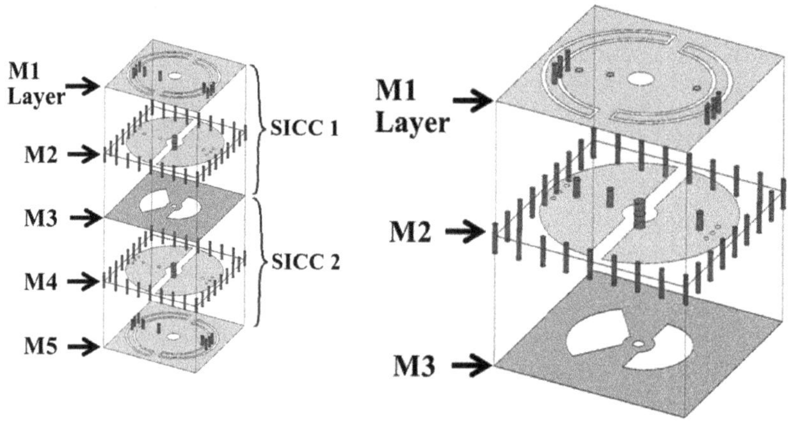

(a) (b)

Figure 4. The proposed extremely miniaturized dual-band SICC BPF structure: (**a**) 3D view and (**b**) zoom-in of SICC unit.

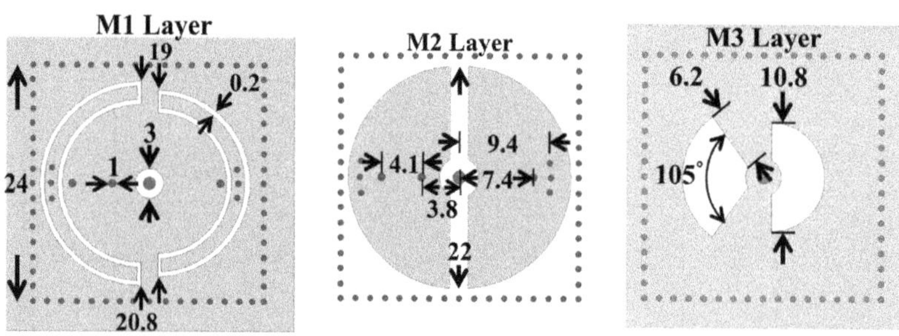

(a) (b) (c)

Figure 5. The layouts and circuit dimensions of the BPF in Figure 4b: (**a**) M1, (**b**) M2, and (**c**) M3 layers.

The coupling between the upper and lower SICCs relies on the two fan-shaped slots embedded in the common wall (the M3 layer in Figures 4a and 5c) of the two SICCs. The left-side slot depicted in Figure 5c is smaller than the right-side one, which is larger; this configuration allows the left-side slot to suitably function for the high-end passband, while the right-side slot operates better at the low-end passband. The nature of this coupling is classified as an electrical coupling, resulting in the patches on either side of the coupling slot exhibiting opposite electric potentials. The calculated coupling coefficients for the low-end (first) and the high-end (second) passbands as functions of the subtended angle (δ) of the coupling slots are presented in Figure 6 for various radius values (r) of fan-shaped slots.

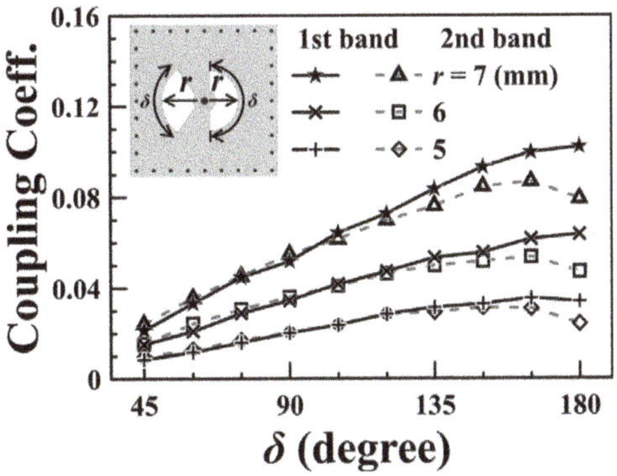

Figure 6. Coupling coefficient versus the slot's subtended angle for various slot radii.

Figure 7 presents the distributions of the electric field (E-field) and the vector magnetic field (H-field) for the first and second passbands. The E-field is calculated on the upper surface of the metal patch, while the H-field is simulated beneath the surface of the metal patch, to which the extra blind shorting via-holes are connected. In the first passband, as illustrated in Figure 7a, the right semicircular patch exhibits a resonance predominating at low frequencies. Furthermore, the H-field is also significantly stronger in the right-side region. In contrast, the second passband reveals a greater E-field strength on the left semicircular patch, accompanied by a more enhanced H-field also in the left cavity. The H-field distribution surrounding the two shorting via-holes of the left patch shows that the H-field encompasses both via-holes. This observation implies that the inductances created by the blind shorting via-holes are arranged in a parallel configuration, resulting in a total inductance less than that of a single blind shorting via-hole. Consequently, this smaller inductance leads to a higher resonance frequency.

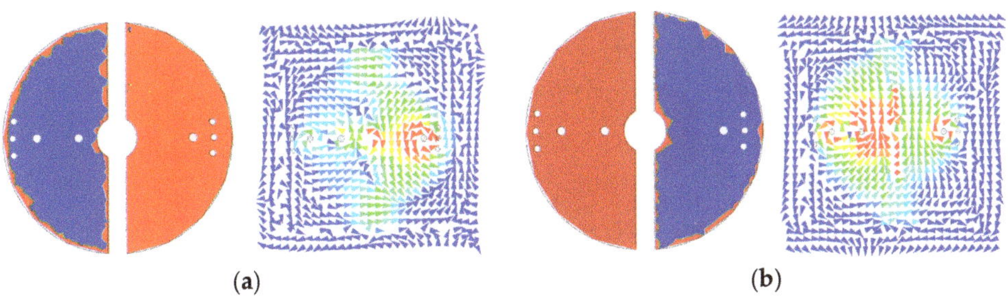

Figure 7. The distributions of electric field and vector magnetic field at the (**a**) first and (**b**) second passbands.

As observed from Figure 7a, when the right-side half-mode coaxial resonator resonates at the first passband, the left-side half-mode coaxial resonator is not entirely silent. Similarly, from Figure 7b, when the left-side half-mode coaxial resonator resonates at the second passband, there are still fields in the right-side half-mode coaxial resonator. This is because the left- and right-side half-mode coaxial resonators are not entirely isolated. Adjustment of the structural dimensions of the left-side (right-side) half-mode coaxial resonator for tuning the lower (higher) resonance frequency would affect the higher (lower) resonance frequency. Hence, we cannot claim that the two resonance frequencies can be independently controlled. Fortunately, the field strengths in the left- and right-side half-mode coaxial resonators are pretty distinct, and structural dimensions can easily be adjusted to obtain the two desired resonance frequencies.

The frequency responses, both measured and simulated, are presented in Figure 8, with the inset illustrating the zoomed-in passband responses. The coupling mechanism is primarily dominated by the electric field (E-field) within the passband regions, transitioning to dominance by the magnetic field (H-field) in the stopband region. A transmission zero (TZ) is generated during the coupling transition from E-field to H-field dominance and vice versa, forming two TZs on either side of each passband. The coupling slot's axial dimensions (and subtended angle δ) are, respectively, 5.4 (δ = 180°) and 6.2 mm (δ = 105°), corresponding to coupling coefficients of 0.042 and 0.038 for the first and second passbands, respectively. The calculated unloaded Q_u values are 166 and 239 for the cavity operating at the first and second resonance frequencies, respectively. These values are lower than the Q_u value of 680 associated with the conventional SIW cavity counterpart, which depletes every structure within the SIW cavity. The prepreg film used in the printed circuit board (PCB) technology for binding purposes significantly impacts the performance of the proposed circuit due to its lossy nature and its placement over the middle patch layer, where the electric field strength is maximized. When depleting the loss associated with the prepreg film, the Q_u value increases to 282 for the first band and 375 for the second band. The measured and simulated performance parameters observed in Figure 8 are summarized in Table 1, which includes mid-band frequencies, 3DB fractional bandwidths (FBWs), minimum in-band insertion losses (ILs), transmission zeros (TZs), and calculated Q_u. The external quality factor, Q_e, is 12.36 for the first band and 25.51 for the second band. Figure 9 shows the photos of the circuit fabricated for measurement. Table 2 compares our design with several previously referenced SIW-related dual-band filters. It should be noted that the upper stopband bandwidths listed in Table 2 are under the criterion of $|S_{21}| \leq -15$ dB. Furthermore, the BPF in [24] has the largest electric dimension in the table and is chosen as the reference for size comparison.

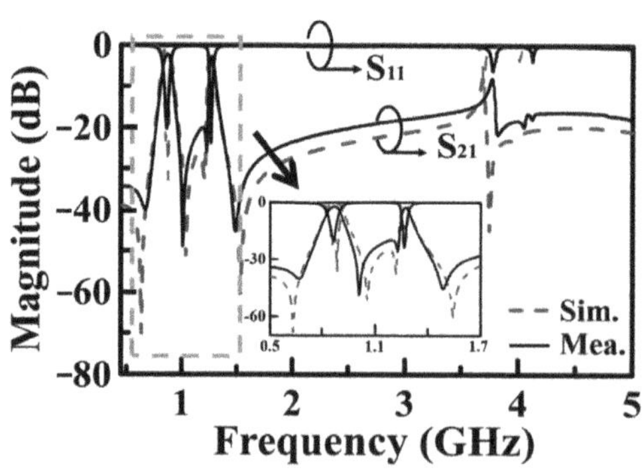

Figure 8. Measured and simulated frequency responses for the BPF of Figure 4.

Table 1. The measured and simulated performance parameters concluded from Figure 8.

	Simulations	Measurements
f_1/f_2 (GHz)	0.87/1.27	0.865/1.275
IL_1/IL_2 (dB)	0.74/1.61	1.94/2.51
3DB FBW_1/FBW_2 (%)	11.49/4.03	8.09/3.92
Stopband BW (GHz) ($|S_{21}| \leq -20$ dB)	1.06	1.76
TZs (GHz)	0.63/1.05/1.22/1.54	0.65/1.01/1.23/1.48
Q_{u1}/Q_{u2}	166/239	

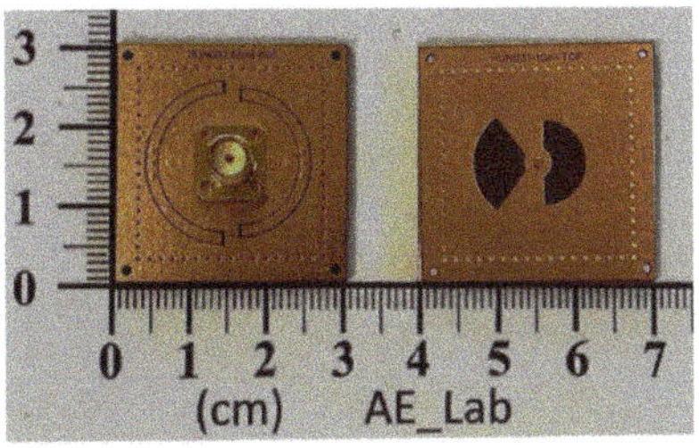

Figure 9. Photos of the experimental circuit, its top and coupling slots views.

Table 2. The comparison between our design and several referenced dual-band SIW cavity BPFs.

	No. of Layers	f_1/f_2 (GHz)	FBW_1/FBW_2 (%)	IL_1/IL_2 (dB)	USB BW	Size ($\lambda_d \times \lambda_d$)	Size Ratio (%)
[10]	2	3.5/5.24	2.73/5.28	1.52/1.65	N/A	1.23 × 1.23	95.3
[23]	3	2.51/5.3	6.8/5.8	1.41/1.88	~0.29f_1	0.84 × 0.42	22.2
[24]	1	7.71/9.64	5.45/8.1	1.9/1.65	N/A	1.26 × 1.26	100
[25]	1	7.89/8.89	3.4/3.9	1.5/1.9	~0.07f_1	1.12 × 1.05	74.1
[26] (Figure 4)	1	3.6/6.4	3.3/2.4	1.3/1.8	>0.72f_1	0.43 × 0.88	23.8
[27] (Figure 6)	1	5.0/7.5	5.46/4.75	1.65/2.25	~0.23f_1	1.65 × 0.93	96.7
[35] (Figure 2)	4	1.81/2.43	6.06/2.9	1.04/2.3	0.24f_1	0.21 × 0.21	2.78
This work	4	0.865/1.275	8.09/3.92	1.94/2.51	2.52f_1	0.103 × 0.103	0.67

USB BW: upper stopband BW with $|S_{21}| < -15$ dB; $\lambda_d = c_0/(f_1\sqrt{\varepsilon_r})$, c_0 is the speed of light.

4. Miniaturized SICC Diplexer

The 3D view of the proposed miniaturized SICC diplexer is presented in Figure 10, while the circuit layouts and dimensions for each of the metal layers are detailed in Figure 11. The diplexer consists of three SICCs, with the common-port (CP) cavity designed to function in a dual-band operation. The SICCs are fabricated using two RT/Duroid 5880 substrates, with thicknesses of 1.57 mm for the upper substrate and 0.254 mm for the lower substrate. Standard PCB technology is utilized for the diplexer's fabrication, employing the same prepreg film referenced in Sections 2 and 3 for binding purposes. The CP cavity incorporates two semicircular patches for dual-band functionality (see Figure 11b). The semicircular patch for the low-band resonance is connected to the cavity's bottom wall through a single extra blind shorting via-hole, and the one for the high-band resonance is equipped with two extra blind shorting via-holes. Per the methodology outlined in Section 2, each half of the CP cavity has a semicircular SCR of varying size embedded in the M1 layer and three via-holes connecting to the patch to facilitate further frequency downshift. Bisecting a single SICC into two equal-sized half-SICCs results in two duplexing-port (DP) cavities. This bisecting is performed by deploying an array, as illustrated in Figure 11d,e, along the symmetric axis of the cavity, thereby physically separating the two DP cavities. Notably, the half-SICC-formed DP cavity cuts the circuit area in half. The high- and low-frequency signals of the CP cavity are coupled to either of the two DP cavities through a semicircular coupling slot for the high-band signal, or two quarter-circular slots for the low-band signal (see Figure 11c). The port-2 cavity in Figure 10c, featuring a larger SCR and a single extra blind shorting via-hole, is designed to respond to low-band resonance. In contrast, the port-3 cavity is equipped with a straight SCR and two extra blind shorting via-holes, thereby serving the high-band function.

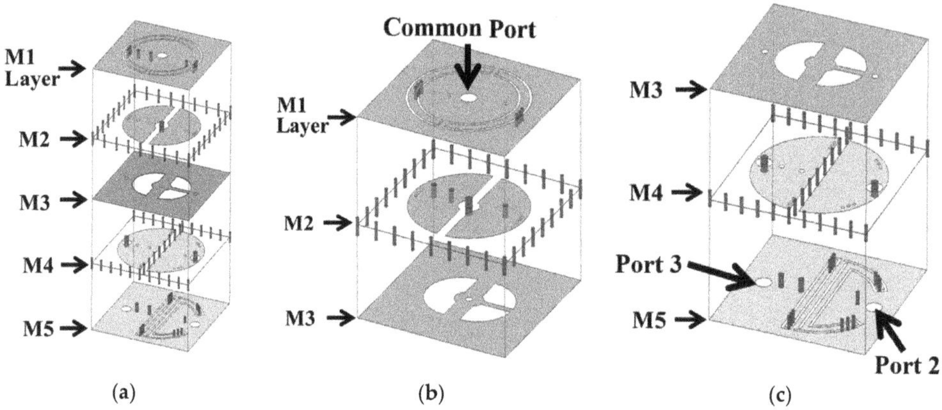

Figure 10. The extremely miniaturized SICC diplexer structure: (**a**) 3D view, (**b**) zoom-in of the common port SICC, and (**c**) zoom-in of the diplexer port cavities.

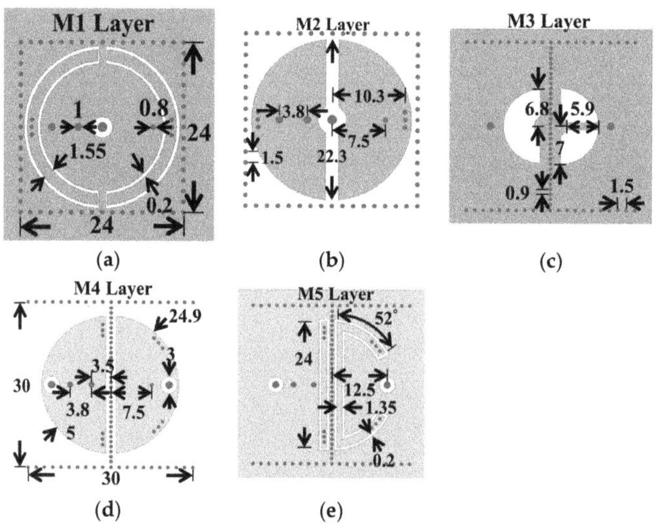

Figure 11. Layouts and circuit dimensions of the metal layers in Figure 10: (**a**) M1, (**b**) M2, (**c**) M3, (**d**) M4, and (**e**) M5 layers.

Figure 12 shows the measured and simulated frequency responses for the diplexer shown in Figure 11. The measured and simulated performance parameters are summarized in Table 3, which includes mid-band frequencies, FBWs, and in-band minimum ILs. Not listed in Table 3, most measured isolations between the two duplexing ports exceed 20 dB. In Figure 13, we give the photos of the circuit fabricated for measurement. Table 4 compares our design with several previously referenced SIW-related diplexers. This table summarizes the number of substrates used in the circuits, mid-band frequencies, minimum ILs, isolation levels between the duplexing bands, and the circuit area, which is defined as the area of the smallest rectangle capable of accommodating the cavities. Here, the diplexer in [29] is chosen as the reference for size comparison. In conclusion, our diplexer demonstrates a significant circuit area efficiency and considerable duplexing-band isolation. However, the ILs are slightly higher than those observed in other designs. These higher ILs might be owing to the significant losses associated with the prepreg film and the conduction losses resulting from the intensified electric field at the SCRs' slot edges.

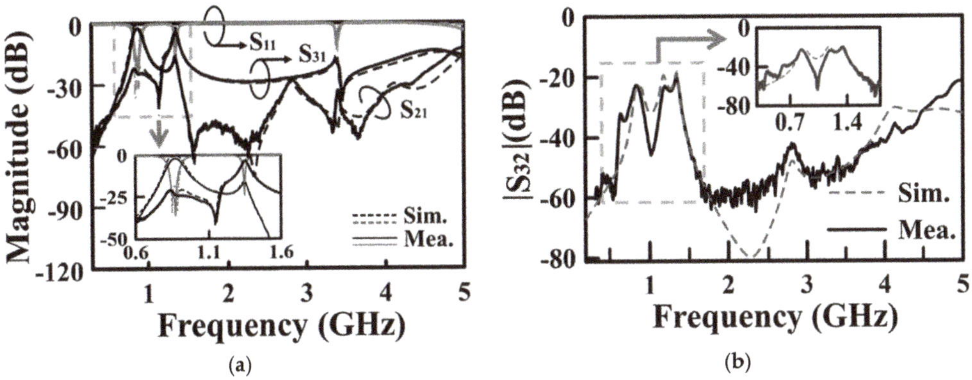

Figure 12. Measured and simulated frequency responses for the SICC diplexer of Figure 11: (**a**) S_{11}, S_{21}, and S_{31}; (**b**) duplexing ports isolation (in terms of $|S_{32}|$).

Table 3. The measured and simulated performance parameters concluded from Figure 12.

	Simulations	**Measurements**
f_1/f_2 (GHz)	0.875/1.35	0.854/1.355
3dB FBW$_1$/FBW$_2$ (%)	12.57/3.27	10.54/3.69
In-band isolations (dB)	19.2–23.4	19.5–23.01
IL$_1$/IL$_2$ (dB)	0.62/2.05	1.63/2.89
Q_{u1}/Q_{u2}	126/275	

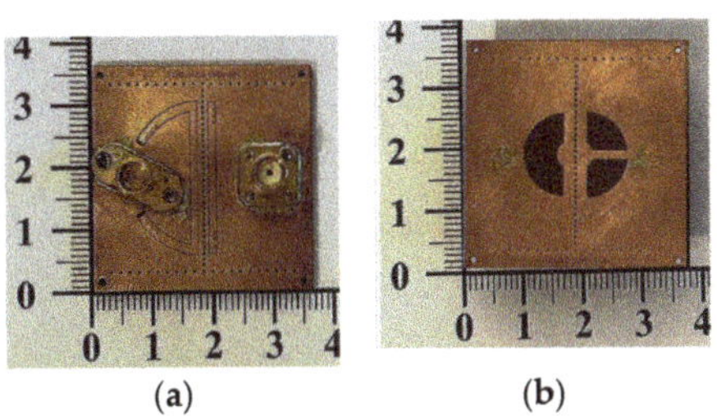

Figure 13. Photos of the experimental circuit in Figure 11: (**a**) duplexing ports and (**b**) coupling slots layer.

Table 4. The comparison of our diplexer design with the previously referenced SIW diplexers.

	No. of Substrates	**f_1/f_2 (GHz)**	**IL_1/IL_2 (dB)**	**IBI (dB)**	**Size ($\lambda_d \times \lambda_d$)**	**Size Ratio (%)**
[28]	1	9.49/9.91	1.42/1.38	N/A	Large	>>100
[29] (Figure 9)	1	9.94/10.95	1.8/2.5	>50	4.1 × 1.57	100
[30] (Figure 10)	3	2.07/2.71	1.1/1.6	>36	0.39 × 0.38	2.3
[31]	1	12/14	1.34/1.41	>27	1.99 × 1.39	43.0
[32]	3	2.35/29.87	0.94/1.32	>22	0.46 × 0.34	2.43
[35] (Figure 13)	4	1.63/1.96	2.5/2.53	>34	0.19 × 0.19	0.56
This work	4	0.854/1.355	1.63/2.89	>19.5	0.127 × 0.127	0.25

IBI: in-band isolation; $\lambda_d = c_0/(f_1\sqrt{\varepsilon_r})$.

5. Conclusions

This paper presents the design, implementation, and verification of a highly miniaturized dual-band substrate-integrated coaxial cavity (SICC) bandpass filter (BPF) and diplexer. The results demonstrate an excellent agreement between the measured and simulated data, affirming the well-structured circuit's effective performance. The proposed dual-band SICC BPF and diplexer occupy circuit areas of only 0.103 $\lambda_d \times 0.103\ \lambda_d$ and 0.127 $\lambda_d \times 0.127\ \lambda_d$, respectively. To the best of the authors' knowledge, the proposed dual-band SICC BPF and diplexer designs have presented the best circuit area efficiency achieved within the dual-band SIW cavity filters and diplexers categories.

Author Contributions: Conceptualization, M.-H.H. and C.-I.G.H.; methodology, M.-H.H., C.-I.G.H. and K.-Y.L.; software, C.-M.H. and C.-F.J.; validation, M.-H.H. and K.-Y.L.; formal analysis, M.-H.H., C.-M.H. and C.-F.J.; investigation, M.-H.H., C.-M.H. and C.-F.J.; resources, M.-H.H. and C.-I.G.H.; data curation, M.-H.H., C.-M.H. and C.-F.J.; writing—original draft preparation, M.-H.H. and C.-I.G.H.; writing—review and editing, M.-H.H. and C.-I.G.H.; visualization, C.-M.H. and C.-F.J.; supervision, M.-H.H. and C.-I.G.H.; project administration, M.-H.H.; funding acquisition, M.-H.H. All authors have read and agreed to the published version of the manuscript.

Funding: This research is funded by the National Science and Technology Council, R.O.C., under the grant contract NSTC 113-2221-E-018-009. The APC, if applicable, is also supported by the same grant.

Institutional Review Board Statement: Not applicable.

Informed Consent Statement: Not applicable.

Data Availability Statement: The data presented in this study are available on request from the corresponding author.

Acknowledgments: We thank the National Science and Technology Council, R.O.C., for the financial support. We would like to thank Shin-Yi PCB, Taiwan, for preparing the experimental circuits.

Conflicts of Interest: Author Chun-Ming Hung was employed by the company Universal Microwave Technology Inc. The remaining authors declare that the research was conducted in the absence of any commercial or financial relationships that could be construed as a potential conflict of interest.

References

1. Piloto, A.; Leahy, K.; Flanick, B.; Zaki, K.A. Waveguide Filters Having a Layered Dielectric Structure. U.S. Patent 5382931, January 1995
2. Uchimura, H.; Takenoshita, T.; Fuji, M. Development of a laminated waveguide. *IEEE Trans. Microw. Theory Tech.* **1998**, *46* 2438–2443. [CrossRef]
3. Deslandes, D.; Wu, K. Integrated microstrip and rectangular waveguide in planar form. *IEEE Microw. Wirel. Compon. Lett.* **2001** *11*, 68–70. [CrossRef]
4. Wang, Y.; Hong, W.; Dong, Y.; Liu, B.; Tang, H.J.; Chen, J.; Yin, X.; Wu, K. Half mode substrate integrated waveguide (HMSIW) bandpass filter. *IEEE Microw. Wirel. Compon. Lett.* **2007**, *17*, 265–267. [CrossRef]
5. Xie, H.-Y.; Wu, B.; Xia, L.; Chen, J.-Z.; Su, T. Miniaturized half-mode fan-shaped SIW filter with extensible order and wide stopband. *IEEE Microw. Wirel. Compon. Lett.* **2020**, *30*, 749–752. [CrossRef]
6. Grigoropoulos, N.; Sanz-Izquierdo, B.; Young, P.R. Substrate integrated folded waveguides (SIFW) and filters. *IEEE Microw. Wirel Compon. Lett.* **2005**, *15*, 829–831. [CrossRef]
7. Jin, C.; Shen, Z. Compact triple-mode filter based on quarter-mode substrate integrated waveguide. *IEEE Trans. Microw. Theory Tech.* **2014**, *62*, 37–45. [CrossRef]
8. Moscato, S.; Tomassoni, C.; Bozzi, M.; Perregrini, L. Quarter-mode cavity filters in substrate integrated waveguide technology *IEEE Trans. Microw. Theory Tech.* **2016**, *64*, 2538–2547. [CrossRef]
9. Delmonte, N.; Bozzi, M.; Perregrini, L.; Tomassoni, C. Cavity resonator filters in shielded quarter-mode substrate integrated waveguide technology. In Proceedings of the 2018 IEEE MTT-S International Microwave Workshop Series on Advanced Materials and Processes for RF and THz Applications (IMWS-AMP), Ann Arbor, MI, USA, 16–18 July 2018; pp. 1–3.
10. Li, P.; Chu, H.; Chen, R.-S. Design of compact bandpass filters using quarter-mode and eight-mode SIW cavities. *IEEE Trans. Compon. Packag. Manuf. Tech.* **2017**, *7*, 956–963. [CrossRef]

11. Kim, P.; Jeong, Y. Compact and wide stopband substrate integrated waveguide bandpass filter using mixed quarter- and one-eighth modes cavities. *IEEE Microw. Wirel. Compon. Lett.* **2020**, *30*, 16–19. [CrossRef]
12. Wang, X.; Zhu, X.-W.; Jiang, Z.; Hao, Z.-C.; Wu, Y.-W.; Hong, W. Analysis of eighth-mode substrate-integrated waveguide cavity and flexible filter design. *IEEE Trans. Microw. Theory Tech.* **2019**, *67*, 2701–2712. [CrossRef]
13. Dong, Y.D.; Yang, T.; Itoh, T. Substrate integrated waveguide loaded by complementary split-ring resonators and its applications to miniaturized waveguide filters. *IEEE Trans. Microw. Theory Tech.* **2009**, *57*, 2211–2223. [CrossRef]
14. Zhang, Q.-L.; Yin, W.-Y.; He, S.; Wu, L.-S. Compact substrate integrated waveguide (SIW) bandpass filter with complementary split-ring resonators (CSRRs). *IEEE Microw. Wirel. Compon. Lett.* **2010**, *20*, 426–428. [CrossRef]
15. Boubakar, H.; Abri, M.; Akram, A.; Benaissa, M.; Ahmad, S. HMSIW miniaturized bandpass filer loaded with two elliptic complementary split-ring resonators for S-band application. In Proceedings of the 2022 International Conference on Emerging Trends in Electrical, Control, and Telecommunication Engineering (ETECTE), Lahore, Pakistan, 2–4 December 2022.
16. Wu, L.-S.; Zhou, L.; Zhou, X.-L.; Yin, W.-Y. Bandpass filter using substrate integrated waveguide cavity loaded with dielectric rod. *IEEE Microw. Wireless Compon. Lett.* **2009**, *19*, 491–493.
17. Li, L.; Wei, Q.-F.; Li, Z.-F.; Yang, J.-X. Compact concave-convex cavity filters using multilayer substrate integrated waveguide. *Electron. Lett.* **2011**, *47*, 500–502. [CrossRef]
18. Saghati, A.P.; Saghati, A.P.; Entesari, K. Ultra-miniature SIW cavity resonators and filters. *IEEE Trans. Microw. Theory Tech.* **2015**, *63*, 4329–4340. [CrossRef]
19. Yang, T.; Ho, K.; Rebeiz, G.M. Compact self-shielded 2–3 GHz high-Q coaxial fixed and tunable filters. *IEEE Trans. Microw. Theory Tech.* **2014**, *62*, 3370–3379. [CrossRef]
20. Ho, M.-H.; Li, J.-C.; Chen, Y.-C. Miniaturized SIW cavity resonator and its application in filter design. *IEEE Microw. Wirel. Compon. Lett.* **2018**, *28*, 651–653. [CrossRef]
21. Sirci, S.; Martínez, J.D.; Boria, V.E. A novel magnetic coupling for miniaturized bandpass filters in embedded coaxial SIW. *Appl. Sci.* **2019**, *9*, 394. [CrossRef]
22. Li, P.; Chu, H.; Zhao, D.; Chen, R.S. Compact dual-band balanced SIW bandpass filter with improved common-mode suppression. *IEEE Microw. Wirel. Compon. Lett.* **2017**, *27*, 347–349. [CrossRef]
23. Li, M.; Chen, C.; Chen, W. Miniaturized dual-band filter using dual-capacitively loaded SIW cavities. *IEEE Microw. Wirel. Compon. Lett.* **2017**, *27*, 344–346. [CrossRef]
24. Azad, A.R.; Mohan, A. Single- and dual-band bandpass filter using a single perturbed SIW circular cavity. *IEEE Microw. Wirel. Compon. Lett.* **2019**, *29*, 201–203. [CrossRef]
25. Zhang, H.; Kang, W.; Wu, W. Dual-band substrate integrated waveguide bandpass filter utilising complementary split-ring resonators. *Electron. Lett.* **2018**, *54*, 85–87. [CrossRef]
26. Zhang, H.; Kang, W.; Wu, W. Miniaturized dual-band SIW filters using E-shaped slotlines with controllable center frequencies. *IEEE Microw. Wireless Compon. Lett.* **2018**, *28*, 311–313. [CrossRef]
27. Zhou, K.; Zhou, C.-X.; Wu, W. Dual-mode characteristics of half-mode SIW rectangular cavity and applications to dual-band filters with widely separated passbands. *IEEE Trans. Microw. Theory Tech.* **2018**, *66*, 4820–4829. [CrossRef]
28. Chongder, P.; Biswas, A. Flexible design procedure for realisation of dual-mode substrate integrated hexagonal-cavity-based diplexer. *IET Microw. Antennas Propag.* **2017**, *11*, 2083–2090. [CrossRef]
29. Khan, A.A.; Mandal, M.K. Design of planar diplexers with improved isolation using the tunable transmission zeros of a dual-mode cavity filter. *IET Microw. Antennas Propag.* **2017**, *11*, 1587–1593. [CrossRef]
30. Hagag, M.F.; Khater, A.; Hickle, M.D.; Peroulis, D. Tunable SIW cavity-based dual-mode diplexers with various single-end and balanced ports. *IEEE Trans. Microw. Theory Tech.* **2018**, *66*, 1238–1248. [CrossRef]
31. Zhou, K.; Zhou, C.-X.; Wu, W. Compact SIW diplexer with flexibly allocated bandwidths using common dual-mode cavity. *IEEE Microw. Wirel. Compon. Lett.* **2018**, *28*, 317–319. [CrossRef]
32. Zheng, S.Y.; Su, Z.L.; Pan, Y.M.; Qamar, Z.; Ho, D. New dual-/tri-band bandpass filters and diplexer with large Frequency ratio. *IEEE Trans. Microw. Theory Tech.* **2018**, *66*, 2978–2992. [CrossRef]
33. Li, D.; Luo, W.; Chen, X.; Liu, Y.; Xu, K.-D.; Chen, Q. Miniaturized dual-/tri-/quad-band bandpass filters using perturbed multimode SIW cavity. *IEEE Trans. Compon. Packag. Manuf. Technol.* **2023**, *13*, 1685–1693. [CrossRef]
34. Ho, M.-H.; Hsu, C.-I.; Tang, K.-H.; Hong, W. Miniaturized bandpass filter design using half mode substrate integrated coaxial resonators. *Micromachines* **2022**, *13*, 389. [CrossRef] [PubMed]
35. Shen, Y.-L.; Hung, J.-T.; Ho, M.-H.; Hsu, C.-I. Miniaturized substrate integrated waveguide cavities in dual-band filter and diplexer design. *IET Microw. Antennas. Propagat.* **2020**, *14*, 428–434. [CrossRef]
36. Ho, M.-H.; Tang, K.-H. Miniaturized SIW cavity tri-band filter design. *IEEE Microw. Wirel. Compon. Lett.* **2020**, *30*, 589–592. [CrossRef]
37. Hsu, C.-I.; Hong, W.; Chen, M.; Tu, W.-C.; Su, G.-W.; Ho, M.-H. Miniaturized quad-band filter design using substrate integrated coaxial cavity. *Micromachines* **2023**, *14*, 347. [CrossRef]

38. Ho, M.-H.; Hung, C.-M.; Hsu, C.-I. Filter design using ultra-miniaturised substrate-integrated coaxial cavity. *IET Microw. Antennas. Propagat.* **2025**, *19*, e70017. [CrossRef]
39. Liu, A.-Q. *RF MEMS Switches and Integrated Switching Circuits Design, Fabrication, and Test*; Springer: New York, NY, USA, 2010; Chapter 4.

sensors

MDPI

Review

Millimeter-Wave Antennas for 5G Wireless Communications: Technologies, Challenges, and Future Trends

Yutao Yang [1], Minmin Mao [1,*], Junran Xu [1], Huan Liu [1], Jianhua Wang [2] and Kaixin Song [1]

[1] College of Electronic Information and Engineering, Hangzhou Dianzi University, Hangzhou 310018, China; 23040435@hdu.edu.cn (Y.Y.); 23040539@hdu.edu.cn (J.X.); hliu66@hdu.edu.cn (H.L.); kxsong@hdu.edu.cn (K.S.)

[2] School of Electronic Engineering, Hangzhou Dianzi University Information Engineering College, Hangzhou 311305, China; wangjianhua@hziee.edu.cn

* Correspondence: mmm@hdu.edu.cn

Abstract: With the rapid evolution of 5G wireless communications, millimeter-wave (mmWave) technology has become a crucial enabler for high-speed, low-latency, and large-scale connectivity. As the critical interface for signal transmission, mmWave antennas directly affect system performance, reliability, and application scope. This paper reviews the current state of mmWave antenna technologies in 5G systems, focusing on antenna types, design considerations, and integration strategies. We discuss how the multiple-input multiple-output (MIMO) architectures and advanced beamforming techniques enhance system capacity and link robustness. State-of-the-art integration methods, such as antenna-in-package (AiP) and chip-level integration, are examined for their importance in achieving compact and high-performance mmWave systems. Material selection and fabrication technologies—including low-loss substrates like polytetrafluoroethylene (PTFE), hydrocarbon-based materials, liquid crystal polymer (LCP), and microwave dielectric ceramics, as well as emerging processes such as low-temperature co-fired ceramics (LTCC), 3D printing, and micro-electro-mechanical systems (MEMS)—are also analyzed. Key challenges include propagation path limitations, power consumption and thermal management in highly integrated systems, cost–performance trade-offs for mass production, and interoperability standardization across vendors. Finally, we outline future research directions, including intelligent beam management, reconfigurable antennas, AI-driven designs, and hybrid mmWave–sub-6 GHz systems, highlighting the vital role of mmWave antennas in shaping next-generation wireless networks.

Keywords: 5G communications; millimeter-wave (mmWave) antennas; MIMO; beamforming; antenna integration technology; low-loss materials and fabrication

1. Introduction

The rapid development of wireless communication technologies, coupled with the increasing prevalence of smart devices, the emergence of novel multimedia applications, and the widespread deployment of Internet of Things (IoT) systems, has led to an unprecedented surge in data traffic [1,2]. This trend necessitates the development of communication technologies that offer low latency and high reliability to support next-generation applications. For example, 5G wireless networks have been designed to address many of the limitations inherent in 4G systems, particularly from an architectural perspective. Moreover, 5G aims to support innovative services that exceed the technical capabilities and capacity requirements of existing networks, thereby revolutionizing key performance metrics such

https://doi.org/10.3390/s25175424

as data rates, latency, massive connectivity, network reliability, and energy efficiency [3]. These enhancements include ultra-low latency (as low as 1 ms) with ultra-high reliability, peak data rates reaching up to 10 Gbps—and potentially as high as 20 Gbps in certain scenarios—as well as a guaranteed minimum data rate of 100 Mbps. Furthermore, 5G is designed to achieve approximately 100 times greater energy efficiency and three times higher spectral efficiency compared to 4G networks [4].

As 5G technology has evolved, the enormous volume of data traffic and the escalating demand for higher communication capacity and signal integrity have led to the near-total occupation of frequency bands ranging from several hundred MHz to several GHz [5]. To meet these growing demands for increased data rates, two primary approaches are typically considered: transitioning to higher frequency bands to expand bandwidth, or enhancing spectral efficiency through the use of advanced modulation schemes or multiplexing techniques such as MIMO technology. Millimeter-wave (mmWave), with wavelengths between 1 mm and 10 mm, corresponds to a spectrum spanning 30 GHz to 300 GHz. Although the physical definition of mmWave strictly refers to the 30 GHz-300 GHz band (wavelength 1–10 mm), the 20 GHz band is often included in the mmWave category in practical applications. The communications industry, such as 3GPP, uniformly categorizes the frequency bands above 24 GHz as 5G mmWave (FR2 band), for example, 24.25–27.5 GHz (n258) and 26.5–29.5 GHz (n257). mmWave has emerged as a critical candidate for 5G wireless systems due to their broad bandwidth and high transmission rates [6,7]. These attributes make mmWave well-suited to fulfill future capacity needs. Currently, mmWave technology is being utilized in various applications, including radar systems (such as automotive radars), military communications, satellite communications, and point-to-point (P2P) communications. Moreover, the adoption of mmWave in 5G networks not only promises to address the increasing bandwidth requirements but also paves the way for innovative services and applications that require ultra-high reliability and low-latency connectivity. This makes mmWave indispensable for realizing the full potential of 5G technology across different sectors.

Antennas, as essential components for transmitting and receiving electromagnetic signals, play a critical role in wireless communication and sensing systems [8]. In the context of mmWave communications, antennas are particularly vital due to the unique propagation characteristics and challenges associated with this high-frequency spectrum. For instance, mmWave signals are highly susceptible to blockage and attenuation from various environmental factors. Rain attenuation is one such challenge—since raindrops are typically on the order of a few millimeters, they can significantly obstruct mmWave signals, whose wavelengths are similarly sized [9]. Other common sources of signal degradation include foliage loss, human body blockage, and material penetration losses. Given these inherent limitations—such as high path loss, limited diffraction capability, and sensitivity to physical obstructions—advanced antenna designs and robust beamforming/beam tracking techniques are required to ensure reliable mmWave communication links [10]. Antenna arrays have emerged as a preferred solution in mmWave systems, enabling the generation of highly directional and steerable beams, thereby improving gain, directivity, and overall link efficiency [11]. Figure 1 presents the obstacles, pros, and design of mmWave antennas and their applications. Figure 1 presents the obstacles, advantages, design, integration, materials, fabrication, typical types, and applications of mmWave antennas.

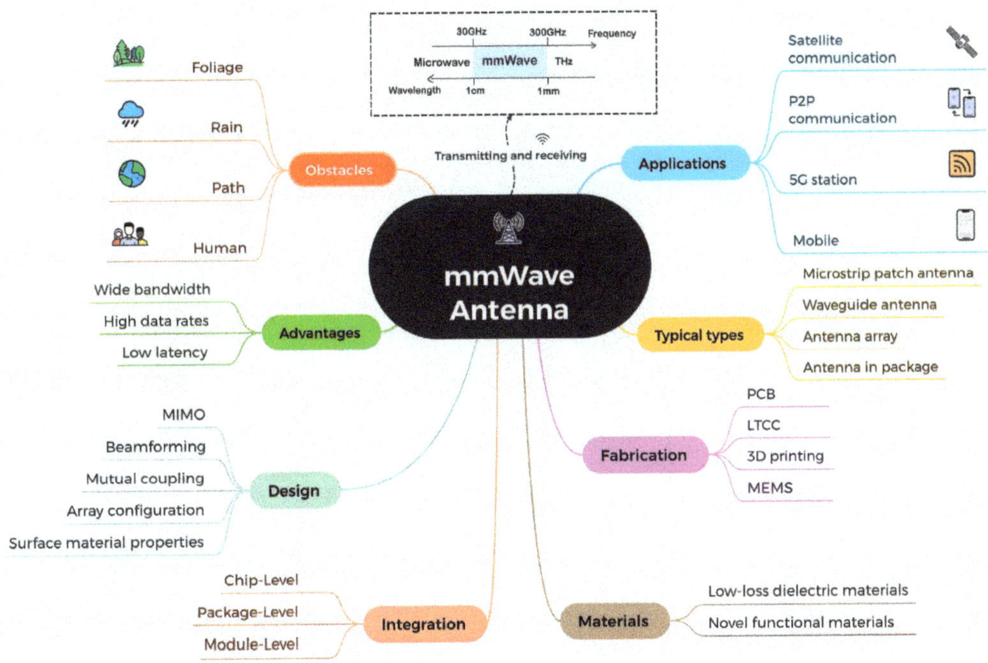

Figure 1. Overview of obstacles, advantages, design, integration, materials, fabrication, typical types, and applications of mmWave antennas.

Against this backdrop, this review article systematically explores the development and challenges of mmWave antennas within the context of 5G wireless communications. We examine key aspects including the types of mmWave antennas, MIMO-based array architectures, performance-influencing factors, integration technologies, as well as materials and fabrication techniques. The structure of this review is organized as follows:

Section 2 provides the methodology for writing this review.

Section 3 introduces the commonly used types of mmWave antennas and their performance characteristics.

Section 4 discusses design considerations for mmWave antennas, with a focus on MIMO architectures and key influencing factors.

Section 5 reviews state-of-the-art antenna integration technologies.

Section 6 presents an analysis of materials and fabrication techniques used in mmWave antenna manufacturing.

Finally, we summarize the current technical challenges and outline future research directions in this rapidly evolving field.

2. Methodology

In this review, we take a systematic approach for a comprehensive and unbiased selection of the relevant work on recent advancements in mmWave antennas for 5G applications. We mainly used Web of Science (WoS) supplemented by IEEE Xplore for conducting a literature search. To ensure the inclusion of the most relevant, highly cited, and up-to-date research, we prioritized peer-reviewed journal articles, review papers, and conference papers published between 2013 and 2025. This time frame was chosen to capture both foundational developments and the latest breakthroughs in the field.

2.1. Databases Searched

The selection of databases is critical for a comprehensive review. This study utilized WoS and IEEE Xplore due to their extensive coverage of peer-reviewed research in electrical engineering, wireless communication, and antenna technologies. WoS provides access to influential multidisciplinary journals, while IEEE Xplore offers a wealth of resources—including journal articles, conference papers, and standards—specifically relevant to mmWave antennas, MIMO systems, and 5G/6G technologies.

2.2. Search Keywords and Boolean Operators

The following keywords were used: mmWave antenna; mmWave MIMO; mmWave decoupling technology; mmWave antenna integration; mmWave materials; mmWave antenna fabrication. Boolean operators were also applied to search the following:

- mmWave antenna OR mmWave MIMO OR mmWave decoupling technology OR mmWave antenna integration OR mmWave materials OR mmWave antenna fabrication.
- mmWave antenna AND mmWave MIMO.

2.3. Inclusion and Exclusion Criteria

We gave priority to the papers that study millimeter wave frequency band and 5G requirements. For reference review articles, we selected the classic papers with high citations. Articles that focused on the design, fabrication, materials, and future trends of mmWave antennas, MIMO, and beamforming techniques were taken into account.

Inclusion Criteria:

- Articles that addressed performance evaluation, challenges, or future trends.
- Studies published between 2013 and 2025.
- Articles focused on design, analysis, fabrication, and materials.
- Classic review papers with high citations (without time window).

Exclusion Criteria:

- Non-English articles.
- Duplicate studies or articles lacking full-text access.
- Research on devices other than antennas.
- Dissertation or Thesis.
- Articles with similar content but published earlier.
- Articles with only abstract.

2.4. Literature Screening Process

After screening, a total of 86 articles were included in the references of this review. Figure 2 presents a summary of the literature screening process, illustrating the number of records identified, screened, and ultimately included in the final analysis.

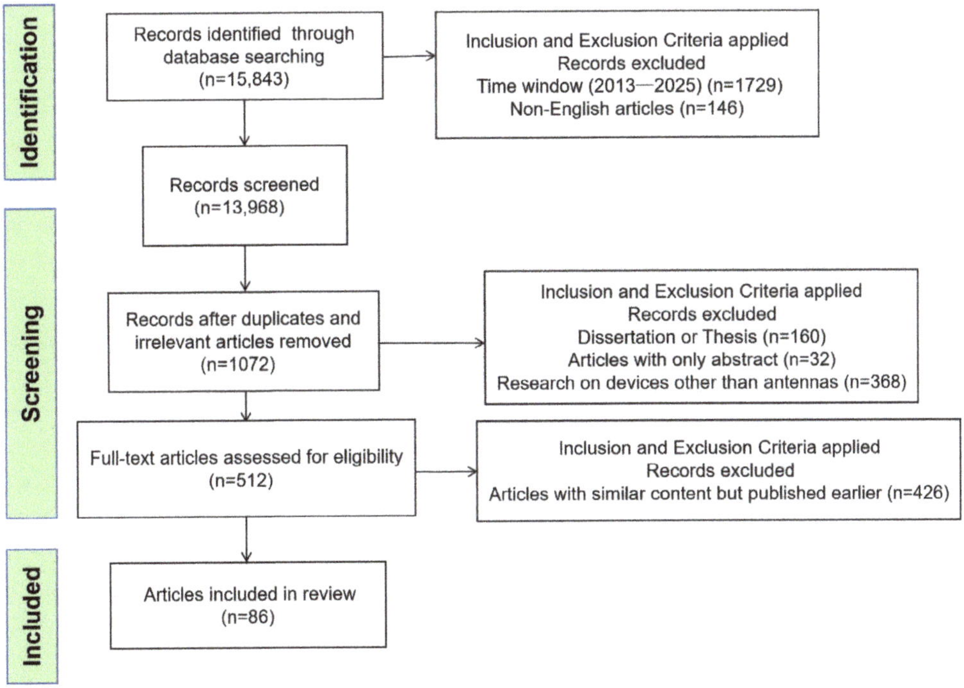

Figure 2. Systematic selection of articles.

3. Typical Types of Millimeter-Wave Antennas

Over the years, several antenna types have emerged as dominant solutions for mmWave applications, each with its own set of advantages, limitations, and suitable use cases. These antennas have been developed and refined in response to the unique challenges posed by mmWave propagation characteristics, such as high free-space path loss, limited penetration through obstacles, and sensitivity to environmental blockage.

3.1. Microstrip Patch Antennas

The microstrip patch antenna was first introduced in the 1970s and is characterized by its use of microstrip lines or coaxial probes to excite the radiating element. Owing to its simple structure, ease of fabrication, and relatively low manufacturing cost, it has been widely adopted in a variety of wireless communication systems [12,13]. With the rapid advancement of mmWave technologies, the compact size and high integration potential of microstrip patch antennas have made them particularly suitable for mmWave applications. Typical implementations include antennas for 5G mobile devices, automotive radar systems for collision avoidance, and Wi-Fi 6E communication modules.

Despite these advantages, microstrip patch antennas also exhibit several inherent limitations that constrain their performance in high-frequency regimes. These include a narrow operational bandwidth, which limits their effectiveness in mmWave scenarios requiring wideband operation; relatively low gain and limited power-handling capability, which restrict their applicability in high-performance systems; and significant mutual coupling in multi-element arrays, often necessitating additional decoupling structures or isolation techniques. Collectively, these challenges represent critical technical barriers that must be addressed to enable further performance enhancement and broader deployment of microstrip patch antennas in next-generation mmWave systems.

To overcome some of these limitations, various structural modifications have been proposed. For example, etching slots into the radiating patch has been shown to improve performance by enabling multimode resonance. The work in [14] describes a symmetric E-shaped patch antenna operating around the Ka-band, formed by introducing a longitudinal slot and two transverse slots on a rectangular patch. The fabricated prototype achieves an impedance bandwidth of 45.4%, with a peak gain of 8.5 dB across the entire operating band and a measured 3 dB gain bandwidth exceeding 30%. Figure 3 presents a compact elliptical slot four-element planar MIMO antenna designed with polarization diversity for mmWave operation in 5G systems, supporting a wide operating bandwidth of 9 GHz (22.2–31.4 GHz) at a −10 dB threshold, demonstrating a maximum gain of 6 dBi and an impressive radiation efficiency of 94%. The antenna's behavior can be readily adjusted by altering the length of the metallic strips integrated into the elliptical slot [15].

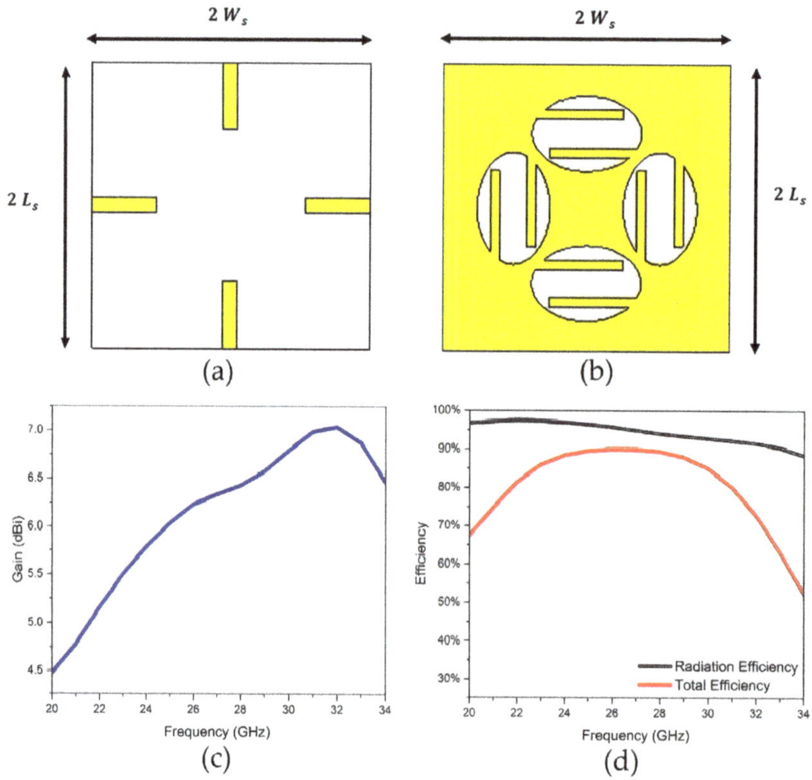

Figure 3. Compact elliptical slot mmWave MIMO antenna: (**a**) front view of antenna section, (**b**) back view of feeding section, (**c**) antenna gain, and (**d**) antenna efficiencies [15].

A square-slot microstrip patch antenna is presented in Figure 4 for mmWave communication at a resonant frequency of 37 GHz [16]. The design incorporates an H-shaped slot on the top side and an inverted T-shaped slot on the bottom of the radiating patch. This optimized structure enhances both the impedance bandwidth and gain. Simulation results show a return loss of −43.05 dB, a gain of 8.245 dB, and an impedance bandwidth of 16.22% at the resonant frequency, as illustrated in Figure 3b.

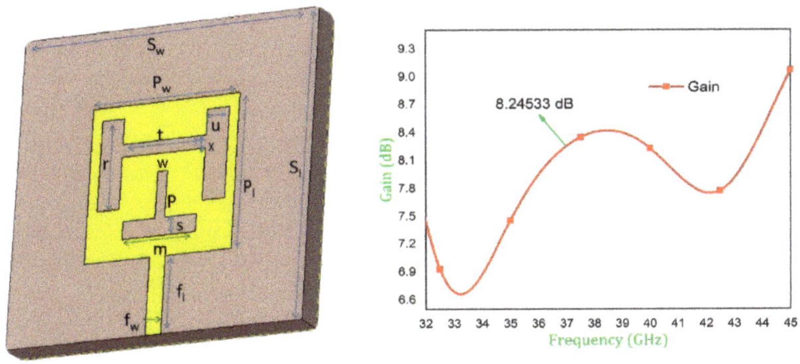

(a) (b)

Figure 4. (**a**) Structure of the proposed antenna and (**b**) the gain of the antenna [16].

In addition to slot-based designs, several studies have explored structural and feeding optimizations to enhance performance. For instance, multilayer configurations with electromagnetic coupling feeds and grid arrays have been proposed to improve gain and bandwidth range [17]; coplanar parasitic elements have been employed to enhance the front-to-back ratio and power-handling capability [18]; and out-of-phase patch arrangements have been utilized to reduce mutual coupling and improve radiation pattern symmetry [19].

A performance comparison of typical mmWave microstrip patch antennas is shown in Table 1. As can be seen from Table 1, most microstrip patch antennas operate at the low-frequency band of mmWave and have the advantage of small size. In addition, the Rogers RT5880 series and Taconic TLY series are popular for the substrate type, showing their good characteristics. Although multi-layer antennas extend the operating frequency band of the antenna, the assembly complexity and interlayer effects are also not negligible. It is expected that the subsequent research on the microstrip patch antenna will make progress on the high-frequency band of mmWave and structure optimization in the future.

Table 1. Performance comparison of mmWave microstrip patch antennas.

Ref.	Antenna Type	Substrate Type	Antenna Dimension	Operating Bands (GHz)	Gain (dBi)	Efficiency	Impedance Bandwidth
[14]	Symmetrical E-shaped	Taconic TLY	0.46 mm × 0.70 mm	37.5	8.5	≥85%	45.4%
[15]	Elliptical Slot	Rogers RT5880LZ	16 mm × 16 mm	28	6	94%	N/A
[16]	Square Slot	Rogers RT5880	12 mm × 12 mm	37	8.18	N/A	16.22%
[17]	Multiband Multilayered	Taconic TLY-5	8.9 mm × 10 mm	40–80	8.65	N/A	N/A
[19]	Linear Antenna Array	Roger 5880	17.45 mm × 99.2 mm	24–31	19.88	86%	5.37 GHz

3.2. Waveguide Antenna

The waveguide antenna is a classical antenna structure that has been widely used in high-performance applications where microstrip patch antennas are less suitable. Compared with microstrip patch antennas, waveguide antennas offer higher power-handling capacity and radiation efficiency, better gain performance, and wider operational bandwidth. With the advancement of mmWave technologies, it has regained significant attention due to its high directivity, excellent radiation efficiency, and low loss characteristics. A typical waveguide antenna consists of three main components: the waveguide body (usually rectangular, circular, or multi-mode), the radiating aperture (e.g., horn-shaped, slot-type, or open-ended), and the feeding structure, which often involves a transition from coaxial or microstrip lines to the waveguide input. The key advantages of waveguide antennas lie in their ability to deliver high gain and strong directional radiation patterns, low insertion loss, and robust mechanical stability, making them well-suited for operation in harsh environments and broadband scenarios. Representative applications include mmWave base station antennas in 5G systems, terahertz (THz) imaging systems employed in medical diagnostics and security screening, and spaceborne communication front-ends operating in the Ka-band and beyond. However, waveguide antennas also face several limitations, including relatively large physical dimensions that hinder miniaturization, high manufacturing complexity leading to increased costs, and integration challenges when interfacing directly with radio frequency (RF) chips, thereby limiting their applicability in compact and highly integrated wireless devices.

To address these challenges, recent studies have proposed various innovative solutions. In order to optimize the antenna structure to reduce the manufacturing complexity and cost, Moschner et al. adopted a novel double-dipole waveguide feed structure based on low-cost additive manufacturing technology, enabling dual-polarized radiation and supporting polarization diversity or MIMO functionality. This approach overcame the limitations of traditional mmWave antennas in terms of cost, fabrication complexity, and performance [20]. In addition, a directional horn antenna was proposed, which adopted the substrate-integrated waveguide (SIW) technology that is superior to microstrip and conventional waveguides in the mmWave region, dual-element arrays, and extended structures to improve the antenna structure and achieve good directional characteristics [21].

Building on leaky-wave concepts, similarly, work in [22] utilized SIW technology for a novel reconfigurable H-plane horn leaky-wave MIMO antenna. This design incorporated dielectric loading and metamaterial structure arrays to boost gain, achieving 7.4 dBi at 24.99 GHz and 8.1 dBi at 26.1 GHz for its 4 × 4 MIMO configuration. A shared-aperture 2D leaky-wave antenna array with polarization and radiation beam reconfigurability was presented in [23]. This design eliminated open-circuit band-stopping through a novel cell featuring two asymmetric slots, attaining a peak gain of 23.6 dBi and 56.6% bandwidth (19–34 GHz).

For high-gain and low-loss array applications, the study [24] presented the design of a high-gain 16 × 16-slot antenna array. Ridge gap waveguide technology was used to reduce the feeding network loss and achieve a low-loss array antenna. The feed layer of the proposed antenna was coupled to a standard rectangular waveguide (WR-28) using a proper transition. The measured results showed an impedance bandwidth of more than 17% over the frequency range of 27.5–32.6 GHz, a maximum gain of 28.9 dBi, and SLL lower than − 20 dB. Zhang et al. adopted the design of all-metal planar array antenna to avoid the serious dielectric loss of mmWave and the combination of ridge gap waveguide (RGW) and E-plane groove gap waveguide (E-GGW) radiation elements to improve the bandwidth [25]. A maximum gain of 27.7 dBi was obtained, with an impedance matching bandwidth of 46.8% during the 18.8 to 30.3 GHz period.

Table 2 shows the research results of several mmWave waveguide antennas, highlighting key performance metrics and design configurations reported in recent years. Among various waveguide technologies, SIW-based antennas have gained increasing popularity and are moving toward mainstream adoption due to high performance and cost-effectiveness. However, the size of SIW-based antennas becomes more challenging at higher mmWave frequencies, where miniaturization is critical. Manufacturing complexity also increases due to tighter tolerances and precision requirements in fabrication. The optimization of waveguide antenna performance therefore demands innovative structural designs. Continued progress will require joint efforts from researchers to address these technical challenges.

Table 2. Performance comparison of waveguide antennas.

Ref.	Antenna Type	Operating Bands (GHz)	Peak Gain (dBi)	Return Loss (dB)	Impedance Bandwidth	Design
[21]	SIW horn	24–28	8.06	−30.89	N/A	Semicircular structure
[22]	SIW H-plane horn leaky wave	22.8 and 26.32	7.3 and 8.1	−33.5	N/A	Dielectric loading and metamaterial structures
[23]	Shared-aperture 2D leaky-wave array	19–34	23.6	N/A	56.6%	A unit cell with two asymmetrical slots
[24]	Slot array	27.5–32.6	28.9	N/A	17%	Ridge gap waveguide and a tapered feeding network
[25]	Full-metal planar array	18.8–30.3	27.7	N/A	46.8%	Double-step double-ridged slot element

3.3. Antenna Array

While antenna arrays have been mentioned in previous sections, this section provides a more detailed and systematic overview of their architecture, functionality, and role in mmWave communications. An antenna array consists of multiple radiating elements arranged in a predefined configuration, where the phase and amplitude of each element are controlled to achieve beamforming. This capability enables enhanced signal strength, improved spatial directionality, and dynamic adaptation to channel conditions, making antenna arrays a cornerstone technology for mmWave communications [26,27]. With the rise of large-scale MIMO systems, antenna arrays have become essential for realizing high-data-rate and low-latency wireless links in next-generation networks. The fundamental components of an antenna array include individual radiating elements—such as patch, waveguide, or Vivaldi antennas [28]—a feeding network (e.g., Butler matrices or T/R modules), a phase-shifting mechanism for beam steering, and a packaging structure that ensures mechanical protection and system integration. Key advantages of antenna arrays include their ability to provide high gain and directional radiation patterns, support intelligent beam management, improve spectral efficiency through spatial multiplexing, and meet the stringent requirements of enhanced mobile broadband (eMBB) and ultra-reliable low-latency communication (URLLC) scenarios. Notable applications include mmWave user equipment (UE), such as the external mmWave antenna module in the Motorola Moto Z3 smartphone; active antenna units (AAUs) in mmWave base stations from vendors like Nokia AirScale and ZTE; and airborne communication systems for high-speed unmanned aerial vehicles (UAVs), where stable mmWave links are critical. Despite their performance benefits, antenna arrays face several technical and practical challenges, including increased system complexity and power consumption, mutual coupling between closely spaced elements that degrades isolation, complicated calibration and synchronization mechanisms, and high manufacturing and deployment costs, which hinder their widespread adoption in cost-sensitive and compact devices. The following describes several promising solutions proposed in recent years.

In 2020, Dai et al. proposed a novel reconfigurable intelligent surface (RIS) comprising 256 2-bit elements (Figure 5), which integrated both phase-shifting and radiation functionalities on a single electromagnetic surface. Operating at 28.5 GHz, the design achieved an antenna gain of 19.1 dB while significantly reducing power consumption compared to conventional phased array antennas [29].

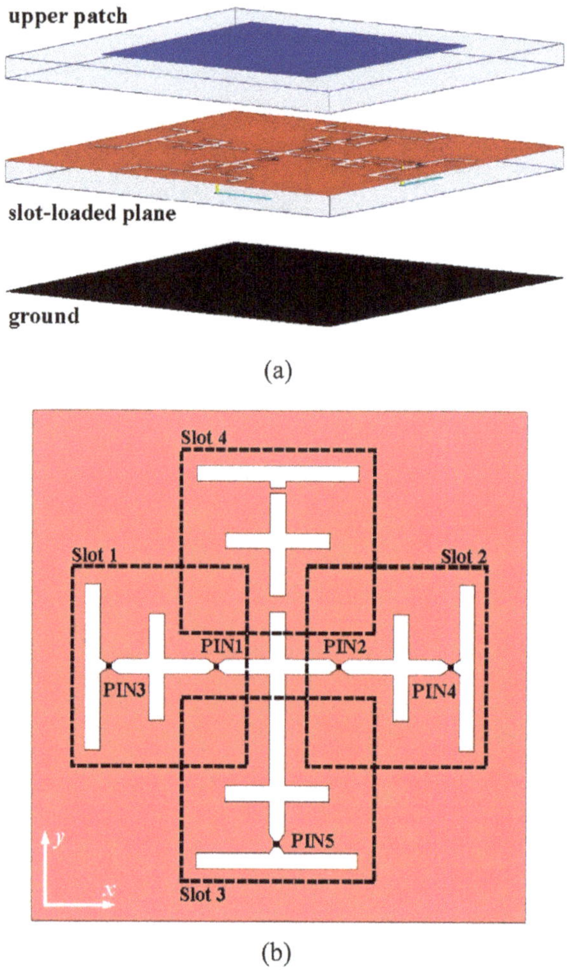

Figure 5. Structure of the proposed 2-bit RIS element: (**a**) exploded view; (**b**) detailed view of the slot-loaded plane [29].

A low-cost antenna array utilizing 3D-printed dielectric polarizers was proposed, in which the entire structure was implemented on a single substrate [30]. This design significantly reduced manufacturing complexity and cost. Furthermore, the adoption of a microstrip line (MSL) feeding mechanism facilitated seamless integration with transceiver circuits, offering a distinct advantage over other waveguide-fed antennas. Table 3 shows the performance comparison of the above antenna arrays.

Table 3. Performance comparison of antenna arrays.

Ref.	Antenna Type	Element Number	Operating Frequency (GHz)	Peak Gain (dBi)	Substrate
[28]	Vivaldi	1 × 4	24.19–29.15, 30.28–40.5	13.2	RT/Duroid 5880
[29]	RIS	256	28.5	19.1	N/A
[30]	Planar	4 × 4	24	20	Rogers 5880 with an MSL feed

In terms of array antenna beamforming synchronization and calibration in time and frequency domains, several algorithmic improvements have been introduced. For instance, a synchronization framework for hybrid mmWave MIMO systems was proposed [31], and the Swift-Link algorithm was also described [32], both aiming to enhance system stability and performance under practical operating conditions. Solutions of beam management based on artificial intelligence (AI) and machine learning (ML) frameworks have also garnered significant attention, particularly in applications such as beam prediction and beam tracking [33]. An AI/ML model can intelligently predict optimal beam directions by analyzing historical channel state information, user behavior patterns, or environmental sensing data, which is very promising in dealing with complex beam situations of large antenna arrays. However, the computational and resource costs of model training remain a notable concern.

3.4. Antenna-in-Package (AiP)

AiP technology integrates antenna elements directly with RF front-end circuits within a single package, offering a compact and system-level solution for mmWave communication systems. This approach addresses the increasing demand for miniaturization, high integration, and cost-effective manufacturing in next-generation wireless devices. An AiP antenna typically consists of radiating elements—often based on patch, slot, or dipole configurations—substrate materials such as organic laminates, LTCC, or fan-out wafer-level packaging (FOWLP), and embedded RF components including phase shifters, power amplifiers, and beamforming ICs. One of the key advantages of AiP antennas is their ability to minimize interconnect losses between the antenna and transceiver, thereby improving overall system efficiency. Additionally, they support multi-band and multi-polarization operations while enabling 3D packaging and heterogeneous integration. These features make AiP antennas particularly suitable for mobile and wearable applications where space is limited but performance requirements are high. Representative commercial implementations include Qualcomm's QTM052 mmWave antenna module for 5G smartphones, Intel's AiP solutions in early 5G laptop modems, and Samsung's phased-array-based AiP modules used in automotive radar and short-range communication systems. Despite these benefits, AiP technology faces several challenges, including thermal management due to the close proximity of active components, limitations in radiation efficiency caused by substrate losses, and difficulties in achieving wide bandwidth and high gain simultaneously. Moreover, standardization and cost-efficient mass production remain critical issues that must be addressed to enable broader deployment across consumer electronics and industrial applications.

Embedded wafer-level ball grid array (eWLB) packaging offers a cost-effective solution for the mass production of AiP [34]. High-density interconnect (HDI) technology provides an alternative for low-cost, large-scale mmWave AiP fabrication. It supports the integration of multiple industry-standard dielectrics with layer thicknesses from 10 to 100 μm, enabling design flexibility and heterogeneous integration. An example is the integrated passive device (IPD) antenna based on a multilayer HDI printed circuit board (PCB) structure presented in [35]. Table 4 presents three processes applied to AiP manufacturing.

Table 4. A comparison table of the three processes.

Type	Dielectric Constant	Loss Tangent	Interconnect Density	Cost
LTCC	5–8	0.003	Low	High
eWLB	3.2	0.004–0.035	High	Low
HDI	3–5	0.003–0.01	Medium	Low

In summary, each type of millimeter-wave antenna possesses distinct advantages and inherent limitations, making the selection of an appropriate antenna design critical for specific application scenarios. Future research should focus on technological innovation and improved cost-effectiveness to enable the broader adoption of mmWave communication technologies in consumer electronics and industrial applications. Moreover, interdisciplinary collaboration will be essential to overcoming existing technical bottlenecks and exploring novel solutions.

4. Millimeter-Wave Antenna Design: The Application of MIMO Technology and Factors Related to Antenna Gain and Efficiency

In wireless communication systems operating in the mmWave band, traditional single-antenna systems struggle to meet the demands for high data rates and reliable connectivity due to significant path loss. As a result, MIMO technology has been widely adopted in mmWave antenna design to enhance system capacity and link performance through techniques such as spatial multiplexing, beamforming, and spatial diversity. At the same time, to achieve high-performance antenna arrays within limited space, it is essential to carefully consider key factors that affect antenna gain and efficiency, including beamforming strategies, mutual coupling effects, array configuration, and surface material properties.

4.1. Application of MIMO Technology in Millimeter-Wave Antenna Systems

Millimeter-wave bands have attracted increasing attention due to their abundant bandwidth resources. However, the inherent increase in free-space path loss at these high frequencies presents new challenges for antenna design. According to Friis transmission equation [36],

$$P_r = G_r G_t \left(\frac{\lambda}{4\pi d} \right)^2 P_t \tag{1}$$

where the powers are in linear scale, d is the TX-RX separation distance, λ is the wavelength and G_t and G_r are the transmit and receive antenna gains, P_t is the transmitted power. The received power P_r decreases with the square of the wavelength λ, meaning that mmWave signals suffer significantly more attenuation than lower-frequency signals in the absence of directional gain compensation. To address these challenges, MIMO technology has become a key enabler in mmWave antenna design. By leveraging spatial multiplexing, beamforming, and diversity gains, MIMO overcomes the limitations of single-antenna systems.

Spatial multiplexing enables the simultaneous transmission of multiple independent data streams, thereby increasing data rates. Spatial diversity improves link reliability by exploiting signal replicas over different propagation paths, reducing the bit error rate without increasing bandwidth or transmit power. Beamforming precisely controls the phase and amplitude of each antenna element to focus signal energy in specific directions, compensating for the high path loss characteristic of mmWave signals and reducing interference [37]. By using multiple transmitting and receiving antennas to simultaneously transmit multiple data streams, MIMO significantly enhances data rates and channel capacity. This technology not only increases system throughput but also improves link quality and reliability [38]. With technological advancements, MIMO has evolved into massive MIMO, significantly enhancing spatial multiplexing gains and improving link robustness. MIMO systems can be broadly categorized into four types, as demonstrated in Figure 6, each tailored to specific application scenarios and performance objectives.

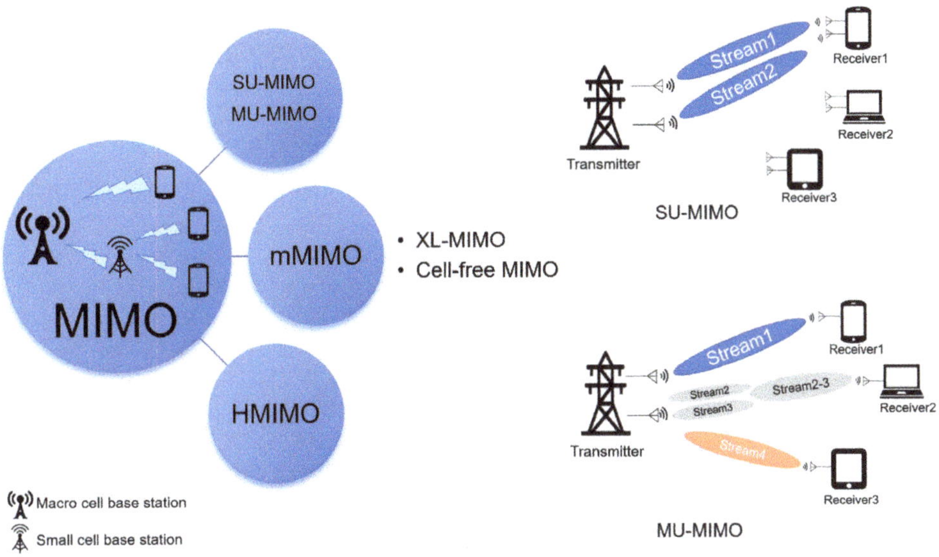

Figure 6. The division of MIMO systems.

4.1.1. Single-User MIMO (SU-MIMO)

SU-MIMO improves spectral efficiency by deploying multiple antennas at both the transmitter and receiver ends of a single user device. It utilizes spatial multiplexing to split data streams into multiple independent streams transmitted simultaneously through different antennas. Spatial diversity is also employed to enhance link reliability without increasing bandwidth or transmit power. At the receiver, multiple antennas are used to separate and reconstruct the original data streams.

4.1.2. Multi-User MIMO (MU-MIMO)

In MU-MIMO systems, a multi-antenna base station serves multiple single-antenna users concurrently. This configuration shifts hardware complexity to the base station side, allowing cost-effective single-antenna terminals while still achieving multiplexing gain. Additionally, MU-MIMO benefits from multi-user diversity, making it more robust than point-to-point SU-MIMO systems in environments with poor scattering conditions [39].

4.1.3. Massive MIMO (mMIMO)

Massive MIMO employs large-scale antenna arrays at the base station to serve multiple users simultaneously, offering significant improvements in spectral efficiency, energy efficiency, and throughput [40–42]. A typical mMIMO system may involve hundreds of antennas serving dozens of users, with each user able to reduce its transmit power proportionally based on the number of base station antennas [39]. mMIMO has become a cornerstone of 5G and is evolving towards extremely large MIMO (XL-MIMO) and cell-free mMIMO architectures in preparation for 6G deployments.

The basic idea of XL-MIMO is to deploy an extremely large number of antennas in a compact space. Compared with mMIMO, XL-MIMO has the following characteristics: more flexible hardware designs, a much larger number of antennas, the much smaller antenna spacing, new EM characteristics, near-field based signal processing, which can provide stronger beamforming gain, richer degrees of freedom (DoF), and spectral efficiency (SE) [43].

Since users at cell boundaries may suffer from strong inter-cell interference, the cell-free MIMO system removes the concept of cell boundaries. By deploying a large number of geographically distributed access points (APs) connected to a central processing unit (CPU), cell-free MIMO can effectively address the inter-cell interference that exists in the intrinsic implementation of a "cell-centric" network [44].

The two technological approaches obtain convenience, but bring greater power consumption, computational complexity, and channel security problems, which necessitate algorithmic and hardware enhancements to meet the above challenges.

4.1.4. Holographic MIMO (HMIMO)

Emerging from advances in metamaterials and RIS, HMIMO represents a paradigm shift by transforming wireless environments into programmable entities. mMIMO uses discrete deployed antennas to form a discrete array aperture, while HMIMO unit elements are placed more and more densely to form an almost spatially continuous aperture, making them capable of forming very sharp beams with weak sidelobes. Interestingly, the mutual coupling effect, which is considered harmful in traditional communication systems, can be used appropriately in HMIMO surfaces to achieve super-directionality [45,46]. Utilizing holographic modulation and continuous aperture surfaces, HMIMO aims to achieve unprecedented spectral efficiency and spatial resolution while maintaining low hardware complexity [47]. It holds great promise for future THz communications, dense IoT networks, and smart radio environments, although it remains in early research stages with many technical challenges yet to be resolved such as complex channel estimation, resource allocation, and beamforming; therefore, there are no unified technical standards and mature industrial ecology, and there is still a long way to go before large-scale commercial deployment.

4.2. Key Factors Affecting Antenna Gain and Efficiency in Millimeter-Wave Systems

The performance of mmWave antennas is influenced by a range of critical factors, including beamforming strategies, mutual coupling effects, array configuration, and surface material properties. These elements directly impact key antenna metrics such as radiation efficiency, directional stability, and overall communication quality. Given the high-frequency nature of mmWave systems and the stringent performance requirements, the influence of these factors becomes even more pronounced. To address these challenges, various optimization approaches and decoupling techniques have been developed and widely adopted. These methods aim to enhance antenna gain, improve beam steering accuracy, and effectively suppress electromagnetic interference between array elements, thereby ensuring reliable and efficient operation in mmWave communication systems.

4.2.1. Beamforming Strategies

Beamforming is considered as a key enabling technique for mmWave band communications [48], which is usually achieved by directing the transmitted signal towards the receiver while suppressing it in a direction other than that of the intended receiver in order to provide significant array gain, providing a higher signal-to-noise ratio (SNR) and additional radio link margin, thus mitigating the propagation path loss, which can be accomplished using digital, analog, or a combination of digital and analog beamforming techniques [37].

Digital beamforming (DBF) relies on pre-processing the transmitted signal in the digital domain and then post-processing the received signal at the receiver [49]. However, digital beamforming adds an extra cost at high frequencies because each antenna requires its own analog RF front-end chain, which results in larger transceivers and higher power consumption.

Analog beamforming is a simple but effective method to bend and control electromagnetic waves without the need for digital signal processing. The technique uses a series of analog devices, including phase shifters and amplifiers, to change the phase and amplitude of the signal in real time to ensure that the signal is transmitted or received in a certain direction or at a specified destination [50], and can produce high beamforming gains from a large number of antennas, but is not as flexible as digital beamforming, and it is this trade-off between flexibility/performance and simplicity that is driving the need for hybrid beamforming architectures, especially when large numbers of antennas are required, such as in the mmWave band [51].

Hybrid beamforming allows the transmitted signal to be processed first through a phase shifter in the digital domain, without the need for an RF chain. As a result, the dimensionality of the signal can be greatly reduced. The post-processed signals undergo conventional analog beamforming to build mmWave MIMO systems with much lower complexity. At the heart of hybrid beamforming is the division of precoding between the analog and digital domains, which allows for an effective trade-off between low-complexity but limited performance analog beamforming and high-complexity, high-performance all-digital precoding [52,53].

4.2.2. Mutual Coupling Effects

In compact MIMO antenna designs, mutual coupling is a significant challenge. To meet miniaturization requirements, the spacing between antenna elements is often reduced, leading to increased electromagnetic coupling, which affects impedance matching and radiation efficiency [54,55]. This effect is even more pronounced in the mmWave band due to shorter wavelengths. To mitigate this issue, various decoupling techniques have been proposed, such as defective ground structures (DGSs) [56], parasitic element decoupling techniques (PDTs) [57], slit structures [58], dielectric resonator antennas (DRAs), vias [59], complementary split-ring resonators (CSRRs) [60], and electromagnetic bandgap (EBG) structures [61]. These techniques aim to provide high isolation while reducing antenna size, thereby improving overall antenna performance. The performance of different decoupling techniques is shown in Table 5.

Table 5. Isolation improvements of different decoupling techniques.

Ref.	Technology	Operating Bands (GHz)	Mutual Coupling Reduction/ Isolation Improvement	Substrate	Relative Design Complexity	Antenna Dimension (mm^2)
[56]	DGS	3–3.6, 3.6–3.9	Isolation improved to more than 10 dB	FR-4	Low	19.5 mm × 7.4 mm
[57]	PDT	2.45	Mutual coupling reduced to around −40 dB	Rogers RO4350	High	N/A
[58]	Slit structures	3.6	High isolation with more than −25 dB mutual coupling	FR-4	Low	25 mm × 25 mm
[59]	DRAs with vias added vertically	25–27	E-plane and H-plane coupling reduced 19.8 and 22.7 dB, leading to a high isolation level of over 30 dB	Rogers 6010	Low	N/A
[60]	CSRRs	36–50	Isolation improved to 22 dB	Rogers RO4350B	Low	28 mm × 28 mm
[61]	EBG structures	28	−60 dB, −72 dB of peak mutual coupling reduction and −25 dB of isolation improvement	Rogers RO4350B	High	N/A

4.2.3. Array Configuration

The configuration of an antenna array significantly impacts the performance of mmWave antennas. Different array configurations (such as rectangular, cross-shaped, circular, and hexagonal arrangements) affect directionality, gain, and main lobe width. For example, circular arrays, due to their larger area and uniform element spacing, offer higher array gain and directivity. In terms of beam alignment, circular arrays have wider main lobes, resulting in less performance loss when the beam is not perfectly aligned. In contrast, narrow beams in rectangular and cross-shaped arrays are more susceptible to antenna vibration or external environmental factors, increasing the likelihood of communication disruptions [10]. Therefore, carefully selecting and designing array configurations is crucial for optimizing the performance of mmWave antennas.

4.2.4. Surface Material Properties

The choice of surface coating materials also has a significant impact on the performance of mmWave antennas. Different surface coating materials (such as Au/Pd/Ni(P) (ENIG), Au/Pd/Ni(P) (ENEPIG), ultra-thin Ni(P)-type ENEPIG, immersion tin (ImSn), Au/Pd(P)/Au (IGEPIG), and immersion gold (IG)) affect the scattering parameters (S-parameters), radiation patterns, gain, and temperature distribution of the antenna. Studies [62] have shown that surface coatings containing thick Ni(P) layers can lead to center frequency shifts and gain degradation, whereas ultra-thin Ni(P)-type ENEPIG and IGEPIG coatings exhibit better antenna performance (Figure 7). Additionally, with the development of metamaterials and RIS, new possibilities for surface coating techniques based on these advanced materials are emerging, providing innovative solutions for mmWave antenna design.

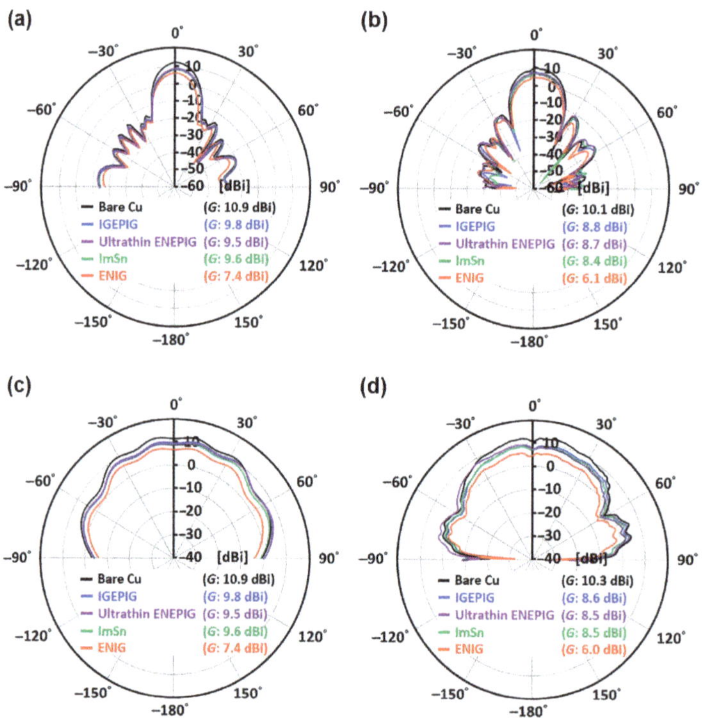

Figure 7. Two-dimensional radiation patterns of the comb-line antenna with different surface finishes at f = 77 GHz: (**a**) HFSS simulation (yz-plane); (**b**) anechoic chamber measurements (yz-plane); (**c**) HFSS simulation (xz-plane); (**d**) anechoic chamber measurements (xz-plane) [62].

In summary, MIMO technology serves as a cornerstone in mmWave antenna design, effectively addressing challenges such as high path loss and strong interference inherent in the mmWave band. It significantly enhances data transmission rates and channel capacity, laying a solid foundation for more efficient and reliable wireless communication systems. In parallel, factors including beamforming strategies, mutual coupling, array configuration, and surface material properties collectively determine the radiation efficiency, directivity, and overall performance of mmWave antennas. The synergistic integration of MIMO technology with these key antenna design considerations is not only essential for maximizing the potential of mmWave communications but also lays a solid technical foundation for the evolution towards 6G and future THz communication systems.

5. Integration Technologies for Millimeter-Wave Antennas

With the increasing demand for high data rates, low latency, and dense connectivity in 5G and future 6G communications, mmWave antenna systems are evolving towards highly integrated architectures [63,64]. This trend not only helps reduce device size and power consumption but also enhances overall system performance and deployability. Currently, the integration technologies of mmWave antennas can be categorized into three main levels: chip-level integration, package-level integration, and module-level integration. The relationship among the three levels is shown in Figure 8. These approaches offer complementary advantages across different application scenarios and together facilitate the commercialization of mmWave communication systems.

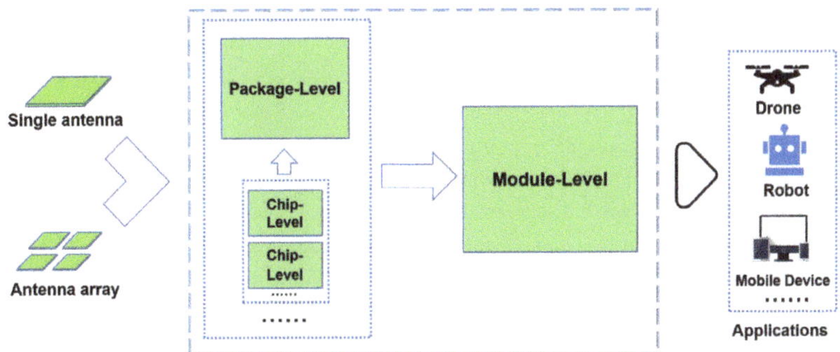

Figure 8. Three levels of integration technologies for mmWave antennas.

5.1. Chip-Level Integration

Chip-level integration refers to the co-design and co-fabrication of antenna elements with RF front-end circuits—such as low-noise amplifiers (LNAs), power amplifiers (PAs), and phase shifters—on the same substrate, or even within a single system-on-chip (SoC) architecture. This approach minimizes interconnect losses and parasitic effects that typically occur between discrete components, making it particularly suitable for compact system designs at high frequencies.

The key advantages include significantly reduced signal path length, which lowers transmission loss and improves efficiency, as well as enabling fine-grained beamforming control for enhanced directivity and stability. However, chip-level integration also presents several challenges, such as the high dielectric loss of silicon substrates at mmWave frequencies, thermal management issues, and potential mutual interference between the antenna and active circuits. As a result, researchers have been exploring alternative high-performance semiconductor materials—such as GaAs, SiGe, and GaN—to optimize the radiation efficiency and overall performance of on-chip antennas [65].

5.2. Package-Level Integration

Package-level integration, commonly implemented as AiP, represents a compromise between chip-level and module-level integration. In this approach, the antenna is embedded within the chip packaging structure, allowing it to coexist with the RF transceiver chip inside the same package. AiP strikes a balance between performance and cost-effectiveness, making it the dominant solution for mmWave applications in mobile devices.

By leveraging advanced packaging technologies—such as FOWLP, flip-chip bonding, and wafer-level packaging (WLP)—AiP enables the integration of multi-antenna arrays and RF front-ends within limited space. Its benefits include good electromagnetic compatibility, ease of mass production, and compatibility with standard CMOS processes. Additionally, AiP supports the design of multi-band and multi-polarization antennas, meeting the diverse requirements of 5G smartphones, AR/VR headsets, automotive radars, and other mobile platforms.

Despite its advantages, AiP faces certain limitations, such as dielectric losses from packaging materials and restricted antenna size that may compromise directivity. Therefore, ongoing research focuses on developing new packaging materials with low dielectric constants, 3D stacked structures, and heterogeneous integration techniques to further enhance AiP antenna performance [66].

5.3. Module-Level Integration

Module-level integration involves integrating the mmWave antenna array, RF front-end modules (RFFEs), beamforming chips, filters, and power management units into a single functional module. This approach retains a degree of flexibility and maintainability while achieving high system integration, making it widely used in base stations, access points, industrial robots, drones, and other semi-fixed or fixed deployment scenarios.

The advantages of module-level integration include rapid deployment through standardized interfaces, simplified system complexity, and improved reliability through optimized thermal and mechanical design. High-performance substrates such as LTCC, HDI PCB, and flexible PCB are often employed, along with metal shielding and cavity filters to achieve excellent electrical performance and interference suppression.

Moreover, module-level integration supports advanced architectures such as large-scale MIMO and hybrid beamforming, providing a viable technical platform for emerging technologies like RIS and distributed antenna systems (DASs) in 6G communications [67,68].

In conclusion, the integration technologies for mmWave antennas are evolving towards a multi-layered, synergistic development path. Chip-level integration emphasizes performance optimization and miniaturization, package-level integration prioritizes process compatibility and mass production feasibility, and module-level integration focuses on system scalability and engineering practicality. Table 6 summarizes the key differences among the three levels of integration technologies. Together, these integration strategies form a critical technical chain—from core components to full system deployment—for realizing mmWave communication systems. With advancements in advanced manufacturing processes, novel materials, and AI-assisted design tools, mmWave antenna integration technologies will continue to evolve, laying a solid foundation for the upcoming era of 6G and THz communications.

Table 6. Three levels of integration technologies.

Integration Level	Operating Frequency (GHz)	Thickness (mm)	Typical Size	Interconnect Loss	Typical Technologies
Chip-level	<200	0.05–1	μm~mm	Low	CMOS, SiGe
Package-level	<120	0.1–1.5	~mm	Medium	eWLB, FOWLP, HDI
Module-level	<60	0.5–3	mm~cm	High	LTCC, PCB HDI

6. Fabrication and Material Selection for Millimeter-Wave Antennas

Millimeter-wave antennas operate in the high-frequency range of 30 GHz to 300 GHz, where signal wavelengths are short and manufacturing precision, material performance, and process compatibility are critically important. The choice of fabrication method and materials directly affects key performance metrics such as radiation efficiency, insertion loss, and directivity, and also determines the feasibility and stability of the antenna in practical communication systems. Currently, common fabrication techniques include PCB printing, LTCC, 3D printing, and MEMS processing, each with its own characteristics and suitable for different application scenarios. At the same time, the use of low-loss dielectric materials and novel functional materials has introduced new possibilities for mmWave antenna design.

6.1. Fabrication Technologies

6.1.1. PCB Printing Technology

PCB printing is one of the most widely used methods for antenna fabrication, especially suitable for microstrip patch antenna designs. This technology is mature, cost-effective, and conducive to mass production, making it ideal for mid-to-low frequency applications. However, at mmWave frequencies, traditional FR4 substrates exhibit significant dielectric and conductor losses, leading to increased signal attenuation and limiting their use in high-performance mmWave systems. Therefore, replacing conventional materials with low-loss high-frequency substrates—such as PTFE, hydrocarbon-based materials, or LCP—is essential for improving the performance of PCB-based antennas.

6.1.2. LTCC Technology

LTCC is a multilayer ceramic integration technique that offers excellent dielectric properties, thermal stability, and mechanical strength, making it well-suited for mmWave applications [69,70]. With LTCC, complex feed networks and antenna structures can be realized in three-dimensional space, effectively reducing interconnect losses and parasitic effects. Additionally, LTCC supports high-density integration, enabling the co-design of RF FEMs and antennas. It is widely used in mmWave radar, 5G base stations, and satellite communications.

6.1.3. Three-Dimensional Printing Technology

With the development of additive manufacturing technologies, 3D printing has shown unique advantages in the fabrication of mmWave antennas. This technique enables the rapid prototyping of highly complex geometries, such as irregular shapes, curved surfaces, lens antennas, and metamaterial structures, surpassing the limitations of traditional fabrication methods [71,72]. Three-dimensional printing also offers flexible design iteration capabilities and relatively low development costs, making it particularly suitable for prototype verification and small-batch customized production. Currently, both metal 3D printing and polymer-based 3D printing combined with metal coating have been applied in mmWave antenna fabrication, with potential for further improvements in resolution and electrical performance.

6.1.4. MEMS Technology

MEMS technology is used to fabricate miniature reconfigurable antennas capable of dynamically adjusting frequency, polarization, or beam direction [73]. These antennas are particularly suitable for smart beamforming and adaptive communication systems. MEMS switches and varactors can be embedded into the antenna structure to achieve functions such as beam steering and frequency switching. Although still in the developmental stage for mmWave applications, MEMS technology shows great potential in miniaturization, low power consumption, and multifunctional integration.

The key features and typical applications of these technologies are summarized in Table 7. In summary, manufacturing technologies for mmWave antennas are evolving from isolated, single-process approaches toward a holistic paradigm of "material–process–function" co-design. Future advancements will depend not only on improved fabrication precision, but increasingly on the integration of multi-scale manufacturing capabilities and intelligent functionalities. This convergence lays a robust foundation for next-generation 6G and THz communication systems.

Table 7. Summary of fabrication technologies for mmWave antennas.

Fabrication Technology	Features	Applications
PCB Printing	Mature and cost-effective Suitable for mass production Typical size accuracy: <0.1 mm Frequency range: up to 100 GHz	Mid-to-low-end mmWave devices 5G terminals Consumer electronics
LTCC	Multilayer integration Supports 3D structures Dielectric constant: 3~10 Thermal expansion coefficient: <10 ppm/°C Bending strength: >100 MPa Elastic modulus: >80 GPa Frequency range: up to 300 GHz	RF FEMs mmWave radar 5G base stations Satellite communications
3D Printing	Enables complex geometries Relatively low development cost Supports rapid prototyping Surface roughness: <50 μm Frequency range: up to 300 GHz	Prototype verification Customized small-batch production Novel antenna structures
MEMS	Miniaturized and reconfigurable Enables dynamic control of frequency Actuation voltage: 5–50 V Response time: μs~ms Frequency range: up to 100 GHz	Smart beamforming systems Reconfigurable antennas Wearable devices

6.2. Material Selection

6.2.1. Low-Loss Dielectric Materials

Millimeter-wave antennas are highly sensitive to the dielectric constant (ε_r) and loss tangent (tanδ) of the substrate materials. Commonly used low-loss high-frequency dielectric materials include the following:

- PTFE: A widely used high-frequency material due to its extremely low dielectric constant ($\varepsilon_r \approx 2.0$–$2.2$) and loss tangent (tanδ ≈ 0.0004–0.001), making it ideal for microwave and mmWave applications. It is often reinforced with glass fibers or other fillers to improve mechanical rigidity. Its chemical inertness and thermal stability further ensure reliable performance under harsh conditions. It offers stable dielectric properties and good thermal management, making it widely used in mmWave antenna applications [74,75].
- Hydrocarbon-based materials: Cost-effective and process-friendly alternatives that combine hydrocarbon resins with ceramic fillers to achieve moderate dielectric constants ($\varepsilon_r \approx 3.0$–$4.0$) and low loss tangent (tanδ ≈ 0.002–0.004). They offer good dimensional and thermal stability, and importantly, they can be manufactured using standard FR-4 processes, which significantly reduces production costs. Their electrical performance is comparable to that of PTFE-based materials.
- LCP: A thermoplastic material characterized by low dielectric loss (tanδ ≈ 0.002–0.004), excellent moisture resistance, and good thermal stability over a wide temperature

range. Its flexibility and dimensional stability make it particularly suitable for flexible and wearable mmWave antenna designs [76,77].

- Microwave dielectric ceramics: Known for their superior electrical and mechanical properties, especially in high-frequency and high-power applications. These ceramics have a wide range of dielectric constants ($\varepsilon_r \approx 5$–40) and extremely low loss tangents ($\tan\delta \approx 0.0001$–0.001), along with high mechanical strength and thermal stability. This makes them ideal for components requiring precise frequency control and long-term reliability, such as filters, resonators, and antennas. Additionally, microwave dielectric ceramics are compatible with LTCC technology, enabling compact integration and widespread use in mmWave radar systems, 5G base stations, and satellite communication modules [78–81].

In fact, dielectric loss is highly frequency dependent. Figure 9 illustrates the variation of the loss tangent for PTFE, hydrocarbon-based materials, LCP, and microwave dielectric ceramics as a function of frequency. Therefore, the operating frequency must be carefully considered when selecting low-loss dielectric materials. The evolution of low-loss dielectric materials is shifting from isolated property optimization toward a system-level paradigm that emphasizes multifunctionality, process compatibility, and environmental robustness.

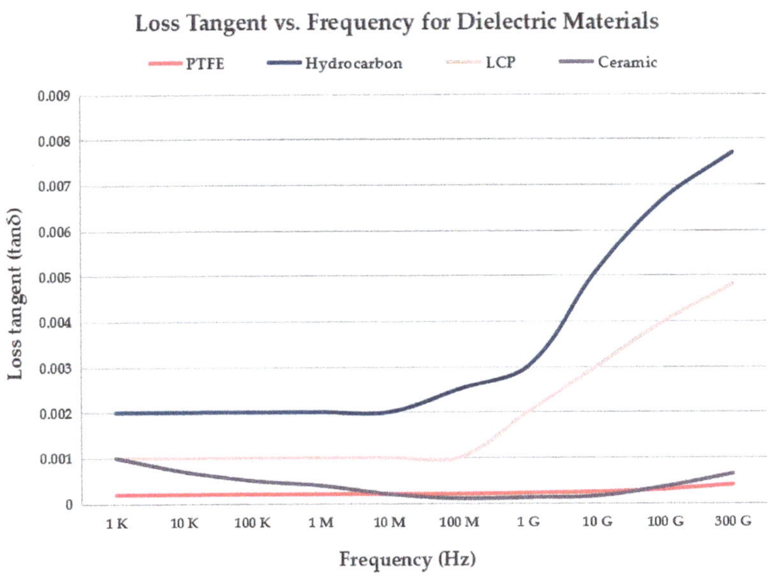

Figure 9. Schematic plot of loss tangent as a function of frequency for low-loss dielectric materials.

6.2.2. Novel Functional Materials

In recent years, several emerging materials have brought breakthroughs to mmWave antenna design:

- Graphene: With ultra-high electron mobility and tunable electromagnetic response [82], graphene can be used in reconfigurable antennas, absorbers, and frequency selective surfaces (FSSs), enhancing antenna flexibility and performance.
- Metamaterials: Engineered sub-wavelength structures enable exotic electromagnetic properties such as negative refraction, perfect absorption, and anomalous reflection, helping realize compact, highly directive, and broadband mmWave antennas [83,84].
- Smart and Phase-Change Materials: Examples are VO_2 (vanadium dioxide) and GST ($Ge_2Sb_2Te_5$), which can switch between metallic and insulating states under exter-

nal stimuli (e.g., current, light, temperature), enabling dynamically tunable antenna functions [85,86].

The selection of these materials should consider multiple factors, including cost, processability, environmental adaptability, and compatibility with system integration requirements. To achieve optimal performance in mmWave antenna design, careful material selection is essential. The representative examples of the above-mentioned materials and their key advantages are summarized in Table 8. Beyond conventional dielectrics, the emergence of novel functional materials is opening transformative pathways for next-generation mmWave antennas. While challenges remain in large-scale fabrication, long-term reliability, and integration complexity, these advanced materials represent a paradigm shift from passive to active, intelligent, and adaptive antenna systems. Their development not only expands the design space for mmWave antennas but also aligns closely with the vision of cognitive radio, smart environments, and 6G wireless networks, where responsiveness and multifunctionality are paramount.

Table 8. Summary of material selection for mmWave antennas.

Materials		**Examples**	**Advantages**
Low-Loss Dielectric Materials	PTFE	RO 3003 (Rogers) RT/Duroid 5880 (Rogers) RF-35 (Taconic)	Stable dielectric properties Good thermal management Compatibility with PCB processes
	Hydrocarbon-based Materials	RO 4350B (Rogers) RO 4350C (Rogers) I-Tera®MT40 (Isola)	Stable dielectric properties Low cost and thermal stability Compatibility with PCB processes
	LCP	Vectra® (Celanese) Zenite® (DuPont) Xydar® (SABIC)	Very low dielectric loss Moisture resistance Excellent flexibility for wearable devices
	Microwave Dielectric Ceramics	Al_2O_3 Mg_2SiO_4 $Mg_2Al_4Si_5O_{18}$	Ultra-low loss tangent High mechanical strength Multilayer integration (LTCC design)
Novel Functional Materials	Graphene	CVD-grown graphene Graphene-based composites	Ultra-high electron mobility Tunable electromagnetic response Supports reconfigurable designs
	Metamaterials	Split-ring resonators Fishnet structures Artificial magnetic conductors	Exotic EM properties (negative refraction, perfect absorption, etc.) Enhances directivity and bandwidth
	Smart and Phase-Change Materials	Vanadium dioxide (VO_2) Germanium antimony telluride (GST, $Ge_2Sb_2Te_5$)	Dynamic tunability via external stimuli Supports programmable RF components and adaptive antenna functions

In summary, the choice of fabrication technology and material selection plays a critical role in determining the performance of mmWave antennas. Regarding various fabrication techniques—such as PCB printing, LTCC, 3D printing, and MEMS—each offers distinct advantages, making it suitable for different application scenarios, including low-cost consumer devices, high-performance communication systems, prototyping of complex structures, and reconfigurable antenna designs, respectively. At the same time, the appropriate selection of low-loss dielectric and novel functional materials not only significantly improves the radiation efficiency and directivity of antennas but also drives the evolution of mmWave antennas towards greater intelligence, efficiency, and flexibility. Looking ahead, with continuous advancements in advanced manufacturing technologies and materials science, mmWave antennas will play an increasingly important role in cutting-edge fields such as 6G, THz communications, intelligent sensing, and the IoT.

7. Future Challenges and Development Prospects of Millimeter-Wave Antennas

Despite significant progress in mmWave antenna technology over the past decade, with promising applications in 5G communications, vehicular networking (V2X), industrial automation, augmented reality (AR/VR), and satellite communications, several key challenges remain before these systems can be widely deployed and commercialized. At the same time, with the rapid development of artificial intelligence, novel materials, and intelligent surfaces, mmWave antennas are expected to witness broad prospects for innovation and performance enhancement.

7.1. Propagation Path Limitations: Signal Blockage and Coverage Constraints

Millimeter-wave signals operate at high frequencies with short wavelengths, offering abundant bandwidth but suffering from poor penetration and diffraction capabilities. Even small obstacles—such as human bodies, walls, or foliage—can cause severe signal attenuation or even complete link failure, limiting the coverage and reliability of mmWave communication systems. Figure 10 shows a schematic diagram of propagation losses for mmWave and sub-6 GHz as a function of distance.

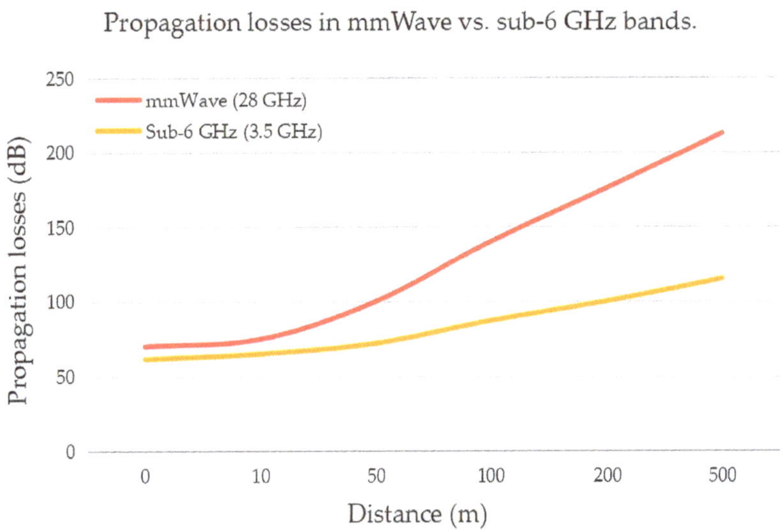

Figure 10. Schematic diagram of propagation losses for mmWave and sub-6 GHz.

To address this issue, current research focuses on the following directions:

- Introduction of RIS: By programmatically controlling the amplitude and phase of electromagnetic waves, RIS can dynamically optimize signal propagation paths, enabling non-line-of-sight (NLoS) transmission and significantly extending coverage.
- Deployment of Relay Nodes and Distributed Antenna Systems (DAS): Strategically placing relay devices in complex environments can establish multi-hop communication links, compensating for coverage gaps caused by line-of-sight limitations.
- Integration with Low-Frequency Bands: Employing a hybrid communication architecture that combines sub-6 GHz bands for robust connectivity with mmWave bands for high data throughput can achieve a balance between performance and coverage.

7.2. Power Consumption and Thermal Management: Heat Dissipation in Highly Integrated Systems

As mmWave antennas evolve towards higher levels of integration—especially in chip-level and package-level integration—RF front-ends, beamforming circuits, and antenna arrays are tightly integrated on the same substrate or within the same package. This results in increased power density and localized hotspots, which threaten device longevity and system stability.

To tackle these thermal challenges, innovative research is being conducted at multiple levels:

- Development of New Thermal Conductive and Dissipative Materials: Materials such as graphene, diamond, and high-conductivity ceramics are being explored to enhance heat transfer efficiency.
- Optimization of Thermal Management Structures: Techniques including microchannel cooling, thermoelectric cooling, and airflow-guiding structures aim to improve overall system thermal performance.
- Low-Power Circuit and Energy-Efficient Algorithm Design: Optimizing RF front-end architectures, reducing beamforming power consumption, and introducing adaptive power management strategies help minimize energy usage.

7.3. Cost and Mass Production Challenges: Balancing Performance and Economics

Millimeter-wave antennas often require high-performance, low-loss materials—such as PTFE, hydrocarbon-based materials, LCP, and microwave dielectric ceramics—as well as precision manufacturing processes like LTCC, MEMS, and high-resolution PCB fabrication. These factors not only increase production costs but also complicate large-scale manufacturing. Moreover, varying performance requirements across different application scenarios and the lack of unified design standards hinder widespread commercial deployment.

Key solutions include the following:

- Exploration of Low-Cost Alternative Materials: Modified polymers and flexible printed substrates offer cost-effective alternatives while maintaining acceptable performance levels.
- Promotion of Standardized Manufacturing Processes: Establishing universal packaging specifications and interface protocols for mmWave antenna modules can facilitate industry-wide collaboration.
- Development of Smart Manufacturing and Automated Testing Technologies: Improving production efficiency, reducing manual involvement, and ensuring product consistency and yield.

7.4. Standardization and Compatibility Issues: Interoperability Across Vendors

Currently, there is no globally unified technical standard for mmWave communications, especially regarding antenna interfaces, beam management protocols, channel models, and RIS control mechanisms. The lack of interoperability among equipment from different vendors hinders global deployment and ecosystem development.

To promote the maturity and adoption of mmWave communications, it is essential to implement the following:

- Establish Unified International Standards: Led by organizations such as 3GPP, IEEE, and ITU, coordinated efforts are needed to advance standardization across all aspects of mmWave communication systems.
- Enhance Cross-Vendor Collaboration and Interoperability Testing: Joint laboratories and open platforms can validate system compatibility and promote convergence.

- Build Open-Source Toolchains and Simulation Platforms: Supporting modeling, algorithm verification, and system evaluation will lower development barriers and accelerate innovation.

7.5. Future Development Directions: Technological Convergence and System Innovation

Looking ahead to the era of 6G and beyond into THz communications, mmWave antennas are expected to evolve towards higher performance, greater intelligence, and more flexibility. Key development trends include the following:

- More Efficient MIMO and Beamforming Algorithms: AI-driven self-learning beam alignment, fast switching, and interference suppression algorithms are becoming mainstream.
- Metamaterials and RIS-Assisted Communications: Metamaterials enable compact broadband antennas, while RIS facilitates dynamic beam control.
- Pre-Research on THz Band Antennas: Early exploration of antenna designs for the 0.1–1 THz band aims to overcome traditional material and process limitations.
- AI-Driven Adaptive Antenna Systems: Deep learning enables real-time environmental sensing and automatic adjustment of antenna parameters to maintain optimal communication states.
- Cost-Effective and High-Stability Manufacturing Processes: Emerging technologies such as flexible electronics, printed electronics, and roll-to-roll manufacturing are being explored for mmWave antenna applications.

8. Conclusions

In summary, mmWave antenna technology plays a pivotal role in advancing 5G wireless communication systems by enabling high-speed, low-latency, and large-scale connectivity. This review has provided a comprehensive overview of the state-of-the-art in mmWave antenna design, covering key aspects such as antenna types, MIMO architectures, and advanced beamforming techniques that significantly enhance system capacity and link robustness. Modern integration strategies—such as AiP and chip-level designs—are instrumental in achieving compact, high-performance modules suitable for mass deployment. The selection of appropriate materials—from low-loss substrates like PTFE, hydrocarbon-based materials, LCP, and microwave dielectric ceramics—together with emerging fabrication technologies such as LTCC, 3D printing, and MEMS—further improves both electrical performance and manufacturing scalability. Despite remarkable progress, several technical challenges remain, including signal propagation limitations, thermal management in highly integrated circuits, cost–performance trade-offs in large-scale production, and the need for standardized interoperability across vendors. However, rapid advancements in novel functional materials—such as graphene, metamaterials, and phase-change materials—combined with AI-driven optimization, smart beam management, reconfigurable antenna architectures, and hybrid mmWave–sub-6 GHz systems—are paving the way for next-generation solutions with enhanced adaptability, efficiency, and deployability. Looking ahead, mmWave technology will not only underpin the evolution of 5G and future 6G networks but also expand into diverse application domains, including smart cities, industrial IoT, and space–air–ground integrated communication systems. Therefore, mmWave antennas are poised to become the critical enablers of ultra-high-capacity, low-latency wireless networks, overcoming the spectrum bottleneck and driving the next wave of global digital transformation through ubiquitous, intelligent connectivity that reshapes industries, cities, and human experiences.

Author Contributions: Conceptualization, Y.Y., M.M. and J.X.; writing—original draft preparation, Y.Y. and M.M.; writing—review and editing, M.M., H.L., J.W. and K.S.; supervision, M.M. and K.S. All authors have read and agreed to the published version of the manuscript.

Funding: This work was supported by the National Natural Science Foundation of China (Grant Nos. 51602147, 52161145401, 51672063), Project of Scientific Research, Technical Exploitation Program of the Guangxi Zhuang Autonomous Region (AD24010021), the Natural Science Foundation of Jiangsu Higher Education Institutions of China (16KJB430017), the Department of Science and Technology of Zhejiang province on "sharp soldiers" and "leading geese" research and development research planning project (No. 2023C01183), the National College Students' Innovation and Entrepreneurship Training Program (No. 202510336035) and the Program of "Xinmiao" (Potential) Talents in Zhejiang Province (No. 2025R407A009).

Institutional Review Board Statement: Not applicable.

Informed Consent Statement: Not applicable.

Conflicts of Interest: The authors declare no conflicts of interest.

References

1. Agiwal, M.; Roy, A.; Saxena, N. Next generation 5G wireless networks: A comprehensive survey. *IEEE Commun. Surv. Tutor.* **2016**, *18*, 1617–1655. [CrossRef]
2. Chettri, L.; Bera, R. A comprehensive survey on Internet of Things (IoT) toward 5G wireless systems. *IEEE Internet Things J.* **2020**, *7*, 16–32. [CrossRef]
3. Shafi, M.; Molisch, A.F.; Smith, P.J.; Haustein, T.; Zhu, P.Y.; De Silva, P.; Tufvesson, F.; Benjebbour, A.; Wunder, G. 5G: A tutorial overview of standards, trials, challenges, deployment, and practice. *IEEE J. Sel. Areas Commun.* **2017**, *35*, 1201–1221. [CrossRef]
4. *ITU-R, M.2083-0*; IMT Vision—Framework and Overall Objectives of the Future Development of IMT for 2020 and Beyond; International Telecommunication Union: Geneva, Switzerland, 2015.
5. Andrews, J.G.; Buzzi, S.; Choi, W.; Hanly, S.V.; Lozano, A.; Soong, A.C.K.; Zhang, J.C. What will 5G be? *IEEE J. Sel. Areas Commun.* **2014**, *32*, 1065–1082. [CrossRef]
6. Rappaport, T.S.; Xing, Y.C.; MacCartney, G.R.; Molisch, A.F.; Mellios, E.; Zhang, J.H. Overview of millimeter wave communications for Fifth-Generation (5G) wireless networks—With a focus on propagation models. *IEEE Trans. Antennas Propag.* **2017**, *65*, 6213–6230. [CrossRef]
7. Rappaport, T.S.; Sun, S.; Mayzus, R.; Zhao, H.; Azar, Y.; Wang, K.; Wong, G.N.; Schulz, J.K.; Samimi, M.; Gutierrez, F. Millimeter wave mobile communications for 5G cellular: It will work! *IEEE Access* **2013**, *1*, 335–449. [CrossRef]
8. He, Y.J.; Chen, Y.L.; Zhang, L.; Wong, S.W.; Chen, Z.N. An overview of terahertz antennas. *China Commun.* **2020**, *17*, 124–165. [CrossRef]
9. Rappaport, T.S.; Gutierrez, F., Jr.; Ben-Dor, E.; Murdock, J.N.; Qiao, Y.J.; Tamir, J.I. Broadband millimeter-wave propagation measurements and models using adaptive-beam antennas for outdoor urban cellular communications. *IEEE Trans. Antennas Propag.* **2013**, *61*, 1850–1859. [CrossRef]
10. Zhang, J.; Ge, X.; Li, Q.; Guizani, M.; Zhang, Y. 5G Millimeter-wave antenna array: Design and challenges. *IEEE Wirel. Commun.* **2017**, *24*, 106–112. [CrossRef]
11. Ghosh, S.; Sen, D. An inclusive survey on array antenna design for millimeter-wave communications. *IEEE Access* **2019**, *7*, 83137–83161. [CrossRef]
12. Wang, Y.; Li, J.J.; Zhu, H.P.; Wang, Q.L.; Sun, T.L.; Ni, T.; Lin, Y.H.; Liu, Y.; Mao, M.M.; Hu, J.; et al. Crystal structure and microwave dielectric property of xMgO-SiO_2 (x = 1~2) system for 5G applications. *Crystals* **2023**, *13*, 1296. [CrossRef]
13. Zhu, S.K.; Huang, Z.C.; Lou, W.C.; Song, K.X.; Khesro, A.; Hussain, F.; Tan, Z.Y.; Luo, X.J.; Mao, M.M.; Xue, L.Y.; et al. 5G microstrip patch antenna and microwave dielectric properties of 4 mol% LiF–MgO–x wt% $MTiO_3$ (M = Ca, Sr) composite ceramics. *J. Mater. Sci. Mater. Electron.* **2021**, *32*, 23880–23888. [CrossRef]
14. Yin, J.X.; Wu, Q.; Yu, C.; Wang, H.M.; Hong, W. Broadband symmetrical E-Shaped patch antenna with multimode resonance for 5G millimeter-wave applications. *IEEE Trans. Antennas Propag.* **2019**, *67*, 4474–4483. [CrossRef]
15. Farooq, N.; Muzaffar, K.; Malik, S.A. Compact elliptical slot millimeter-wave MIMO antenna for 5G applications. *J. Infrared Millim. Terahz Waves* **2024**, *45*, 765–788. [CrossRef]
16. Shamim, S.M.; Dina, U.S.; Arafin, N.; Sultana, S. Design of efficient 37 GHz millimeter wave microstrip patch antenna for 5G mobile application. *Plasmonics* **2021**, *16*, 1417–1425. [CrossRef]

17. Firdausi, A.; Alaydrus, M. Designing multiband multilayered microstrip antenna for mmWave applications. In Proceedings of the 2016 International Conference on Radar, Antenna, Microwave, Electronics, and Telecommunications (ICRAMET), Jakarta, Indonesia, 3–5 October 2016; pp. 99–102.
18. Choudhary, S.D. Design of microstrip rectangular patch antenna using coplanar parasitic Rod elements with Two-Layer substrate coupled integrated feeding line technique. *Wirel. Pers. Commun.* **2023**, *131*, 3073–3087. [CrossRef]
19. Khalily, M.; Tafazolli, R.; Xiao, P.; Kishk, A.A. Broadband mm-Wave microstrip array antenna with improved radiation characteristics for different 5G applications. *IEEE Trans. Antennas Propag.* **2018**, *66*, 4641–4647. [CrossRef]
20. Moschner, O.; Kocabasa, I.; Tebar, A.V.; Pirlet, F.; Kronberger, R. Low-cost additively manufactured high-gain millimeter-wave parabolic reflector antenna with waveguide double-dipole Feed: Millimeter waves meet additive manufacturing. *IEEE Microw. Mag.* **2024**, *25*, 128–136. [CrossRef]
21. Hong, T.; Zheng, S.L.; Liu, R.K.; Zhao, W.T. Design of mmWave directional antenna for enhanced 5G broadcasting coverage. *Sensors* **2021**, *21*, 746. [CrossRef]
22. Tewari, N.; Joshi, N.; Srivastava, S. A novel reconfigurable H-plane horn leaky wave substrate integrated waveguide MIMO antenna for K band. *AEU-Int. J. Electron. Commun.* **2023**, *170*, 154832. [CrossRef]
23. Chen, Y.L.; Zhang, L.; He, Y.J.; Chen, Z.N. A polarization and radiation beam reconfigurable integrated antenna with broadband and high gain for mmWave vehicular communication. *IEEE Trans. Veh. Technol.* **2025**, *74*, 4526–4538. [CrossRef]
24. Zarifi, D.; Saber, A.S.; Zaman, A.U. A high-gain gap waveguide-based 16 × 16 slot antenna array with low sidelobe level for mmwave applications. *Sci. Rep.* **2024**, *14*, 31458. [CrossRef] [PubMed]
25. Zhang, T.L.; Tang, R.; Chen, L.; Yang, S.Y.; Liu, X.S.; Yang, J. Ultra-wideband full-metal planar array antenna with a combination of ridge gap waveguide and E-Plane groove gap waveguide. *IEEE Trans. Antennas Propag.* **2022**, *70*, 8051–8058. [CrossRef]
26. Hong, W.; Choi, J.; Park, D.; Kim, M.S.; You, C.; Jung, D.; Park, J. mmWave 5G NR cellular handset prototype featuring optically invisible beamforming antenna-on-display. *IEEE Commun. Mag.* **2020**, *58*, 54–60. [CrossRef]
27. Caudill, D.; Chuang, A.; Jun, S.Y.; Gentile, C.; Golmie, N. Real-Time mmWave channel sounding through switched beamforming with 3-D dual-polarized phased-array antennas. *IEEE Trans. Microw. Theory Tech.* **2021**, *69*, 5021–5032. [CrossRef]
28. Dixit, A.S.; Kumar, S.; Urooj, S.; Malibari, A. A highly compact antipodal Vivaldi antenna array for 5G millimeter wave Applications. *Sensors* **2021**, *21*, 2360. [CrossRef]
29. Dai, L.L.; Wang, B.C.; Wang, M.; Yang, X.; Tan, J.B.; Bi, S.K.S.; Xu, S.H.; Yang, F.; Chen, Z.; Di, R.M.; et al. Reconfigurable intelligent surface-based wireless communications: Antenna design, prototyping, and experimental results. *IEEE Access* **2020**, *8*, 45913–45923. [CrossRef]
30. Al-Alem, Y.; Sifat, S.M.M.; Antar, Y.M.M.; Kishk, A.A. Millimeter-wave planar antenna array augmented with a low-cost 3D printed dielectric polarizer for sensing and internet of things (IoT) applications. *Sci. Rep.* **2023**, *13*, 9646. [CrossRef] [PubMed]
31. Rodríguez-Fernández, J. Joint synchronization and compressive channel estimation for frequency-selective hybrid mmWave MIMO systems. *IEEE Trans. Wirel. Commun.* **2022**, *21*, 548–562. [CrossRef]
32. Myers, N.J.; Mezghani, A.; Heath, R.W. Swift-link: A compressive beam alignment algorithm for practical mmWave radios. *IEEE Trans. Signal Process.* **2019**, *67*, 1104–1119. [CrossRef]
33. Xue, Q.; Guo, J.J.; Zhou, B.G.; Xu, Y.J.; Li, Z.D.; Ma, S.D. AI/ML for beam management in 5G-Advanced: A standardization perspective. *IEEE Veh. Technol. Mag.* **2024**, *19*, 64–72. [CrossRef]
34. Alhenawy, M.; Schneider, M. Antenna-in-package (AiP) in mm-wave band. *Int. J. Microw. Wirel. Technolog.* **2013**, *5*, 55–64. [CrossRef]
35. Khiabani, N.; Chiang, C.W.; Liu, N.C.; Chen, P.Y.; Kuan, Y.C.; Wu, C.T.M. Metamaterial-enabled ultrawideband mmWave antenna-in-package using heterogeneously-integrated Silicon IPD and HDI-PCB for B5G/6G applications. *IEEE J. Emerg. Sel. Top. Circuits Syst.* **2024**, *10*, 436–453. [CrossRef]
36. Rappaport, T.S. *Wireless Communications: Principles and Practice*, 2nd ed.; Pearson Education India: Noida, India, 2001.
37. Heath, R.W., Jr.; González-Prelcic, N.; Rangan, S.; Roh, W.; Sayeed, A.M. An overview of signal processing techniques for millimeter wave MIMO Systems. *IEEE J. Sel. Top. Signal Process.* **2016**, *10*, 436–453. [CrossRef]
38. Kumar, A.; Ansari, A.Q.; Kanaujia, B.K.; Kishor, J. A novel ITI-shaped isolation structure placed between two-port CPW-fed dual-band MIMO antenna for high isolation. *AEU-Int. J. Electron. Commun.* **2019**, *104*, 35–43. [CrossRef]
39. Lu, L.; Li, G.Y.; Swindlehurst, A.L.; Ashikhmin, A.; Zhang, R. An overview of massive MIMO: Benefits and challenges. *IEEE J. Sel. Top. Signal Process.* **2014**, *8*, 742–758. [CrossRef]
40. Larsson, E.G.; Edfors, O.; Tufvesson, F.; Marzetta, T.L. Massive MIMO for next generation wireless systems. *IEEE Commun. Mag.* **2014**, *52*, 186–195. [CrossRef]
41. Borges, D.; Montezuma, P.; Dinis, R.; Beko, M. Massive MIMO techniques for 5G and beyond—Opportunities and challenges. *Electronics* **2021**, *10*, 1667. [CrossRef]
42. Jain, A.; Yadav, S.K. Design and analysis of compact 108 element multimode antenna array for massive MIMO base station. *Prog. Electromagn. Res. C* **2016**, *61*, 179–184. [CrossRef]

43. Wang, Z.; Zhang, J.Y.; Du, H.Y.; Niyato, D.; Cui, S.G.; Ai, B.; Debbah, M.; Letaief, K.B.; Poor, H.V. A tutorial on Extremely Large-Scale MIMO for 6G: Fundamentals, signal processing, and applications. *IEEE Commun. Surv. Tutor.* **2024**, *26*, 1560–1605. [CrossRef]
44. Liu, Z.L.; Zhang, J.Y.; Liu, Z.H.; Du, H.Y.; Wang, Z.; Niyato, D.; Guizani, M.; Ai, B. Cell-Free XL-MIMO meets Multi-Agent reinforcement learning: Architectures, challenges, and future directions. *IEEE Wirel. Commun.* **2024**, *31*, 155–162. [CrossRef]
45. An, J.C.; Xu, C.; Ng, D.W.K.; Alexandropoulos, G.C.; Huang, C.; Yuen, C.; Hanzo, L. Stacked intelligent metasurfaces for efficient holographic MIMO communications in 6G. *IEEE J. Sel. Areas Commun.* **2023**, *41*, 2380–2396. [CrossRef]
46. Gong, T.R.; Gavriilidis, P.; Ji, R.; Huang, C.W.; Alexandropoulos, G.C.; Wei, L.; Zhang, Z.Y.; Debbah, M.; Poor, H.V.; Yuen, C. Holographic MIMO communications: Theoretical foundations, enabling technologies, and future directions. *IEEE Commun. Surv. Tutor.* **2024**, *26*, 196–257. [CrossRef]
47. Pizzo, A.; Marzetta, T.L.; Sanguinetti, L. Spatially-stationary model for holographic MIMO small-scale fading. *IEEE J. Sel. Areas Commun.* **2020**, *38*, 1964–1979. [CrossRef]
48. Boccardi, F.; Heath, R.W.; Lozano, A.; Marzetta, T.L.; Popovski, P. Five disruptive technology directions for 5G. *IEEE Commun. Mag.* **2014**, *52*, 74–80. [CrossRef]
49. Sohrabi, F.; Yu, W. Hybrid digital and analog beamforming design for large-scale antenna arrays. *IEEE J. Sel. Top. Signal Process.* **2016**, *10*, 501–513. [CrossRef]
50. Arora, A.; Tsinos, C.G.; Shankar, M.R.B.; Chatzinotas, S.; Ottersten, B. Analog beamforming with antenna selection for large-scale antenna arrays. In Proceedings of the ICASSP 2021–2021 IEEE International Conference on Acoustics Speech and Signal Processing (ICASSP), Toronto, ON, Canada, 6–11 June 2021; pp. 4795–4799.
51. Roh, W.; Seol, J.Y.; Park, J.; Lee, B.; Lee, J.; Kim, Y.; Cho, J.; Cheun, K.; Aryanfar, F. Millimeter-wave beamforming as an enabling technology for 5G cellular communications: Theoretical feasibility and prototype results. *IEEE Commun. Mag.* **2014**, *52*, 106–113. [CrossRef]
52. El Ayach, O.; Rajagopal, S.; Abu-Surra, S.; Pi, Z.Y.; Heath, R.W. Spatially sparse precoding in millimeter wave MIMO systems. *IEEE Trans. Wirel. Commun.* **2014**, *13*, 1499–1513. [CrossRef]
53. Chen, C.E. An iterative hybrid transceiver design algorithm for millimeter wave MIMO systems. *IEEE Wirel. Commun. Lett.* **2015**, *4*, 285–288. [CrossRef]
54. Khan, M.A.; Al Harbi, A.G.; Kiani, S.H.; Nordin, A.N.; Munir, M.E.; Saeed, S.I.; Iqbal, J.; Ali, E.M.; Alibakhshikenari, M.; Dalarsson, M. mmWave four-element MIMO antenna for future 5G systems. *Appl. Sci.* **2022**, *12*, 4280. [CrossRef]
55. Khan, D.; Ahmad, A.; Choi, D.Y. Dual-band 5G MIMO antenna with enhanced coupling reduction using metamaterials. *Sci. Rep.* **2024**, *14*, 96. [CrossRef]
56. Abubakar, H.S.; Zhao, Z.Q.; Kiani, S.H.; Khan, S.; Ali, T.; Bashir, M.A.; Cengiz, K.; Kamal, M.M. Eight element MIMO antenna for sub 6 GHz 5G cellular devices. *Phys. Scr.* **2024**, *99*, 085559. [CrossRef]
57. Li, M.; Cheung, S. A Novel calculation-based parasitic decoupling technique for increasing isolation in multiple-element MIMO antenna arrays. *IEEE Trans. Veh. Technol.* **2021**, *70*, 446–458. [CrossRef]
58. Parchin, N.O.; Al-Yasir, Y.I.A.; Ali, A.H.; Elfergani, I.; Noras, J.M.; Rodriguez, J.; Abd-Alhameed, R.A. Eight-element dual-polarized MIMO slot antenna system for 5G smartphone applications. *IEEE Access* **2019**, *7*, 15612–15622. [CrossRef]
59. Pan, Y.M.; Qin, X.; Sun, Y.X.; Zheng, S.Y. A simple decoupling method for 5G millimeter-wave MIMO dielectric resonator antennas. *IEEE Trans. Antennas Propag.* **2019**, *67*, 2224–2234. [CrossRef]
60. Khan, S.; Khan, O.; Shah, S.A.A.; Nasir, J.; Malik, B.T.; Khan, S.; Koziel, S. Highly compact wideband high-gain four-element MIMO antenna for 5G new radio IoT. *IEEE IoT J.* **2025**, *12*, 22350–22365. [CrossRef]
61. Oladeinde, A.K.; Aryafar, E.; Pejcinovic, B. MmWave Tx-Rx self-interference suppression through a high impedance surface stacked EBG. *Electronics* **2024**, *13*, 3067. [CrossRef]
62. Chiang, Y.C.; Chang, Y.H.; Yang, Z.Y.; Huang, C.C.; Ho, C.E. Pronounced effect of surface coating on the antenna characteristics at mmWave frequency band. *Surf. Coat. Technol.* **2025**, *497*, 131801. [CrossRef]
63. Lau, I.; Ekpo, S.; Zafar, M.; Ijaz, M.; Gibson, A. Hybrid mmWave-Li-Fi 5G architecture for reconfigurable variable latency and data rate communications. *IEEE Access* **2023**, *11*, 42850–42861. [CrossRef]
64. Liu, R.; Yu, G.D.; Yuan, J.T.; Li, G.Y. Resource management for millimeter-wave ultra-reliable and low-latency communications. *IEEE Trans. Commun.* **2021**, *69*, 1094–1108. [CrossRef]
65. Pavlidis, S.; Medwig, G.; Thomas, M. Ultrawide-bandgap semiconductors for high-frequency devices. *IEEE Microw. Mag.* **2024**, *25*, 68–79. [CrossRef]
66. Lee, S.Y.; Lee, D.; Zhang, Y.P.; Hong, W.; Ghalichechian, N. History and latest progress in antenna packaging technology: Part 2: Emerging materials and solutions. *IEEE Antennas Propag. Mag.* **2025**, *67*, 2–14. [CrossRef]
67. Gadiel, G.M.; Ibwe, K.; Abdalla, A.T. Energy efficient phase interpolator based hybrid beamforming architecture for massive MIMO system. *Telecommun. Syst.* **2024**, *85*, 1–10. [CrossRef]

68. Chen, J.-C. Energy-efficient hybrid beamforming design for intelligent reflecting surface-assisted mmWave massive MU-MISO systems. *IEEE Trans. Green Commun. Netw.* **2024**, *8*, 330–344. [CrossRef]
69. Matin, M.A. Review on millimeter wave antennas-potential candidate for 5G enabled applications. *Adv. Electromagn.* **2016**, *5*, 98–105. [CrossRef]
70. Li, S.; Li, C.; Mao, M.M.; Song, K.X.; Iqbal, Y.; Khesro, A.; Faouri, S.S.; Lu, Z.L.; Liu, B.; Sun, S.K.; et al. High Q × f values of Zn-Ni co-modified $LiMg_{0.9}Zn_{0.1}$-xNi_xPO_4 microwave dielectric ceramics for 5G/6G LTCC modules. *J. Eur. Ceram. Soc.* **2022**, *42*, 5684–5690. [CrossRef]
71. Park, Y.G.; Yun, I.; Chung, W.G.; Park, W.; Lee, D.H.; Park, J.U. High-resolution 3D printing for electronics. *Adv. Sci.* **2022**, *9*, 2104623. [CrossRef] [PubMed]
72. Wu, G.B.; Chan, K.F.; Shum, K.M.; Chan, C.H. Millimeter-wave and terahertz OAM discrete-lens antennas for 5G and beyond. *IEEE Commun. Mag.* **2022**, *60*, 34–39. [CrossRef]
73. Anagnostou, D.E.; Zheng, G.Z.; Chryssomallis, M.T.; Lyke, J.C.; Ponchak, G.E.; Papapolymerou, J.; Christodoulou, C.G. Design, fabrication, and measurements of an RF-MEMS-based self-similar reconfigurable antenna. *IEEE Trans. Antennas Propag.* **2006**, *54*, 422–432. [CrossRef]
74. Islam, S.; Pham, V.L.; Jang, T.H.; Yoo, H. Wave manipulation with mmwave wide bandwidth and extensive spatial coverage using 1-bit reconfigurable intelligent surface. *Prog. Electromagn. Res.* **2024**, *179*, 83–94. [CrossRef]
75. Mandloi, M.S.; Malviya, L. Dual horn MIMO antenna using SIW technology for mmWave high speed vehicular communication. *Phys. Scr.* **2025**, *100*, 035523. [CrossRef]
76. Hwang, I.J.; Oh, J.I.; Jo, H.W.; Kim, K.S.; Yu, J.W.; Lee, D.J. 28 GHz and 38 GHz dual-band vertically stacked dipole antennas on flexible liquid crystal polymer substrates for millimeter-wave 5G cellular handsets. *IEEE Trans. Antennas Propag.* **2022**, *70*, 3223–3236. [CrossRef]
77. Jilani, S.F.; Munoz, M.O.; Abbasi, Q.H.; Alomainy, A. Millimeter-wave liquid crystal polymer based conformal antenna array for 5G applications. *IEEE Antennas Wirel. Propag. Lett.* **2019**, *18*, 84–88. [CrossRef]
78. Liu, H.; Liu, Y.; Yu, X.Q.; Mao, M.M.; Liu, B.; Bafrooei, H.B.; Li, A.H.; Zhang, Y.Y.; Taheri-Nassaj, E.; Song, K.X. A high-performance, temperature-stable $Mg_{1.99}Ga_{0.01}Si_{0.99}Al_{0.01}O_4$-$CaTiO_3$ microwave dielectric ceramic and its 5G/6G waveguide filter. *J. Eur. Ceram. Soc.* **2023**, *43*, 7471–7477. [CrossRef]
79. Liu, Y.; Mao, M.M.; Ni, T.; Liu, H.; Fang, J.; Li, L.; Bafrooei, H.B.; Shi, F.; Hussain, F.; Wang, D.W. Unleashing the potential of Mg_2SiO_4-based ceramics for millimeter-wave applications: Achieving ultra-low loss with enhanced temperature stability through heterovalent ion substitution. *Ceram. Int.* **2024**, *50*, 329–339. [CrossRef]
80. Fang, J.; Mao, M.M.; She, Y.X.; Yang, Y.T.; Ren, Y.J.; Bafrooei, H.B.; Feizpour, M.; Korotkevich, A.; Leontev, V.S.; Kuz'min, M.P.; et al. Phase stabilization strategy for robust high Q × f values in $MgSiO_3$-based ceramics for millimeter-wave applications. *J. Eur. Ceram. Soc.* **2025**, *45*, 117586. [CrossRef]
81. Xiu, Z.Y.; Mao, M.M.; Lu, Z.L.; Huang, Z.; Qi, Z.M.; Bafrooei, H.B.; Zhou, T.; Wang, D.W.; Lin, H.X.; Taheri-nassaj, E. High-Qf value and temperature stable Zn^{2+}-Mn^{4+} cooperated modified cordierite-based microwave and millimeter-wave dielectric ceramics. *J. Eur. Ceram. Soc.* **2022**, *42*, 5712–5717. [CrossRef]
82. Akbari, M.; Khan, M.W.A.; Hasani, M.; Björninen, T.; Sydänheimo, L.; Ukkonen, L. Fabrication and characterization of graphene antenna for low-cost and environmentally friendly RFID tags. *IEEE Antennas Wirel. Propag. Lett.* **2015**, *15*, 1569–1572. [CrossRef]
83. Hao, H.L.; Hui, D.; Lau, D. Material advancement in technological development for the 5G wireless communications. *Nanotechnol. Rev.* **2020**, *9*, 683–699. [CrossRef]
84. Dixit, A.S.; Kumar, S. A survey of performance enhancement techniques of antipodal Vivaldi antenna. *IEEE Access* **2020**, *8*, 45774–45796. [CrossRef]
85. Teeslink, T.S.; Torres, D.; Ebel, J.L.; Sepulveda, N.; Anagnostou, D.E. Reconfigurable bowtie antenna using metal-insulator transition in vanadium dioxide. *IEEE Antennas Wirel. Propag. Lett.* **2015**, *14*, 1381–1384. [CrossRef]
86. Cao, T.; Wang, R.Z.; Simpson, R.E.; Li, G.X. Photonic Ge-Sb-Te phasechange metamaterials and their applications. *Prog. Quantum Electron.* **2020**, *74*, 100299. [CrossRef]

MDPI AG
Grosspeteranlage 5
4052 Basel
Switzerland
Tel.: +41 61 683 77 34

Sensors Editorial Office
E-mail: sensors@mdpi.com
www.mdpi.com/journal/sensors